DATE			

Introduction

to

Numerical Analysis

Jones and Bartlett Books in Computer Science and Related Areas

Bernstein, A.J., and Lewis, P.M., *Concurrency in Programming and Database Systems*

Causey, R.L., *Logic, Sets, and Recursion*

Chandy, K.M., and Taylor, S., *An Introduction to Parallel Programming*

Flynn, M.J., *Computer Architecture: Pipeline and Parallel System Design*

Gregory, J., and Redmond, D., *Introduction to Numerical Analysis*

Hein, J.L., *Discrete Structures, Logic, and Computability*

Lee, E.S., *Algorithms and Data Structures in Computer Engineering*

Nevison, C.H., Hyde, D.C., Schneider, G.M., and Tymann, P., *Laboratories for Parallel Computing*

Walker, H.M., *The Limits of Computing*

Introduction to Numerical Analysis

John Gregory
Don Redmond

Southern Illinois University at Carbondale

Jones and Bartlett Publishers
Boston London

Editorial, Sales, and Customer Service Offices

Jones and Bartlett Publishers
One Exeter Plaza
Boston, MA 02116
1-617-859-3900
1-800-832-0034

Jones and Bartlett Publishers International
P. O. Box 1498
London W6 7RS
England

Library of Congress Cataloging-in-Publication Data
(not available at time of printing)

Acquisitions Editor: Carl Hesler, Jr.
Manufacturing Buyer: Dana Cerrito
Assistant Production Editor: Nadine Fitzwilliam
Cover Designer: Hannus Design Associates
Printing and Binding: Braun-Brumfield

Printed in the United States of America
98 97 96 95 94 10 9 8 7 6 5 4 3 2 1

To Virginia and Charlotte

Table of Contents

Preface

If we are asked, "What is the most important modern day discipline?", shouldn't most of us say, "Mathematics"? Surely, it is indeed the queen of sciences, uniquely combining beautiful theory and understanding with practical methods, applications and results. Without these combinations we would have little progress in disciplines such as the sciences or engineering and we would not have the computer tools of modern day civilization.

What if we are next asked, "What are the important areas of modern day mathematics?" Our experience is that numerical analysis would be very far down on most people's lists. Why is this? Why is it that few people really know of and understand the combined beauty and practicality of this subject?

A *major purpose* of this book is to convince the reader that numerical analysis belongs very near the top of any such list; that with few prerequisites and minimal effort the reader will be able to combine understanding and practicality in very important and almost unique ways. Specifically, with little more than the basics of calculus we will be able to produce and understand the techniques without which few modern scientific results would be possible.

While techniques and methods are very important, they are not the whole story. We have too often seen our colleagues and students do outrageous things or get very poor results because they didn't understand the basic ideas associated with the methods they were using. One of the first author's earliest industrial memories is of an experienced scientist at Hughes Aircraft Company (a Ph.D.

no less) who wasted several hours of computer time trying to solve a nonlinear equation by Newton's method. The equation was not complicated and could have been done in milliseconds with proper technique. That was over 30 years ago, but even today students who take a "methods" course in numerical analysis encounter the same difficulties.

This book provides a user-friendly introduction to numerical analysis. Our writing style is clear and as informal as possible. We have included many figures and numerical tables to help the reader understand the discussion. Unlike most other books, we have also included problems within the text. These problems are important in that they emphasize or reinforce ideas just covered or they touch on closely related ideas and material.

We are also concerned with the art of numerical analysis. Thus, in discussing different methods we often explain why they work or don't work for specific examples. We will also contrast methods, listing reasons why one method is better or worse than another.

This book is also very flexible. It is structured so that it can be used as the textbook for a course in numerical analysis. The authors have covered Chapters 1–7 and parts of Chapter 8 in one semester. Alternately, it can be used by a reader who wants to learn this material on his/her own.

The prerequisite for this book is one year of calculus, but more experience is always desirable. We suggest the students have some programming experience but it is not an absolute requirement since our numerical tables provide the results and insight usually gained in running computer programs.

In addition to the problems within the text, there are extensive exercises at the end of each chapter. Some provide additional insight, others are computational, others theoretical and some are computing exercises.

Finally, we acknowledge the authors' wives, Virginia Gregory and Charlotte Keller, respectively, for their help in editing this book, Charlie Gibson for typesetting and illustrating this manuscript and Carl Hessler of Jones and Bartlett Publishers for his support in this project.

1 *Things That Can Go Wrong and Sometimes Do*

In this chapter we consider some difficulties which are encountered when we work with real numbers or use exact formulas on a computer. The problem is that we can not usually represent numbers exactly nor can computations be performed precisely. Most of the time these computational problems will be insignificant and things will work out nicely. However, there are special examples of which the reader should be aware.

In the first section, we consider a hypothetical computer which has few place values to represent numbers. It is seen that even simple procedures such as the associativity law for addition or a reasonable approximation of log 2 are no longer possible. In the second section we consider examples of catastrophic cancellation and how to correct them. In the third section we illustrate the concept of ill-conditioned problems. We note that the problems in Section 1.1 can be essentially corrected by using a computer with more significant places. However, in the case of catastrophic cancellation or ill-conditioned problems, new formulas or procedures must be used.

These difficulties are caused by two types of errors. The first type, illustrated in Section 1.1, is *round-off error*, the error in representing real numbers on a computer which has only a finite number of place values. The second type, illustrated in the other two sections, is *truncation error*, the error due to the method or algorithm used.

1.1 The Curse of Finiteness

When doing numerical calculations, we are in some sense doomed from the start. The real numbers are made up of both the rational and irrational numbers. However, our computer numbers are only a finite set of rational numbers with a finite representation. Thus, irrational numbers such as $\sqrt{2}$, π and e are not computer numbers nor are most of the rational numbers. For example, our computer numbers contain a smallest positive (rational) number although neither the real numbers nor the rational numbers contain a smallest positive number.

To oversimplify somewhat, suppose we have a very small computer that computes binary numbers N of the form

$$N = \pm .\, x_1 x_2 x_3 x_4 \pm E\,, \tag{1.1}$$

where $x_1 = 0$ if and only if $N = 0$ while x_2, x_3, x_4 and E can take on the values 0 or 1. This gives us a total of 49 numbers, of which only 16 can be used to fill up the interval $(0, 1)$.

Problem 1.1: Justify the numbers 16 and 49 in the last sentence.

To illustrate problems which may occur, with a concrete example, let us consider the infinite series

$$S = \sum_{n=1}^{+\infty} \frac{(-1)^{n+1}}{n} = 1 - \frac{1}{2} + \frac{1}{3} - \frac{1}{4} + \cdots \tag{1.2}$$

This infinite series is an alternating series whose terms decrease monotonically to zero and so it converges to some real number, which the reader will recall is log 2. With our little hypothetical computer above, we run into problems quickly trying to calculate this sum. We have, in binary arithmetic, that $1 = 0.1000 + E1$, $\frac{1}{2} = 0.1000 + E0$ and $\frac{1}{4} = 0.1000 - E1$. However, the binary expansion of $\frac{1}{3}$ is

$$\frac{1}{3} = \frac{0}{2} + \frac{1}{2^2} + \frac{0}{2^3} + \frac{1}{2^4} + \cdots$$

and it does not terminate. Thus, we must approximate by rounding-off, which is what all computers do. In this case our computer has a choice between $0.1011 - E1$ or $0.1010 - E1$ for the value of $\frac{1}{3}$. Assume we round down and choose the latter. Now, what do we do with $\frac{1}{5}$? Because of the way our computer is set up, the smallest positive computer number we can get is $0.1000 - E1$, which is $\frac{1}{4}$. Thus, to our computer $\frac{1}{5}$ and the remaining terms of S are treated as zeros and the computer value of S in (1.2) is S' where

$$\begin{aligned} S' &= 0.1000{+}E1 \ - \ 0.1000{+}E0 \ + \ 0.1010{-}E1 \ - \ 0.1000{-}E1 \\ &= 1 - \frac{1}{2} + \frac{5}{16} - \frac{1}{4} = \frac{9}{16} = 0.5625 \end{aligned}$$

which is much less than the correct eight place value of 0.69314718 for $\log 2$.

For the rest of this chapter we shall consider some of the consequences of the fact that any computer is only a finite machine. We will find that many of the rules that were true in arithmetic no longer hold on a computer. We give two short examples.

Example 1.1: Addition of zero.

In usual arithmetic, $0 + 0 = 0$. But if we take $a = 0.3 \times 10^{-9}$, which would print out as zero on a machine holding nine significant decimal places, then we find that

$$a + a = 0.6 \times 10^{-9},$$

which would print out as 0.000000001, because it would get rounded up.

Example 1.2: Associativity of addition.

Similarly, in usual arithmetic $(a+b)+c = a+(b+c)$ for any real numbers a, b and c. Not so to a computer. Take $a = 0.1 \times 10$, $b = 0.3 \times 10^{-9}$ and $c = 0.3 \times 10^{-9}$. Then, on the same machine as in Example 1.1, b and c look like zero. Thus

$$(a+b)+c = 1\,,$$

whereas

$$a+(b+c) = 1.000000001\,.$$

The following sections will consider more serious computing problems which are not significantly corrected by using a computer with more significant places.

1.2 Catastrophic Cancellation

Catastrophic cancellation occurs when we subtract two nearly equal numbers. More precisely, we mean that the most significant digits of the difference come from the least significant digits of the numbers being differenced. What happens is that many digits cancel each other out and then when the answer is normalized, that is, the decimal point is moved to the left, meaningless zeros are moved in on the right.

A simple example is the following:

Example 1.3: Let $D = \sqrt{10002} - \sqrt{10001}$.

A computer or calculator which carries 10 significant digits gives a machine value of $D = 5.0 \times 10^{-3}$ since it calculates $\sqrt{10002} = 100.0099995$ and $\sqrt{10001} = 100.0049999$. We see that many of the leading significant digits of the two numbers have been cancelled and that the final digit has just been dropped. Thus we should avoid subtraction of these numbers if possible. This can be achieved by the process of rationalization. Recalling, from algebra, that $a^2 - b^2 =$

$(a-b)(a+b)$, we have

$$\sqrt{x}-\sqrt{y}=\frac{x-y}{\sqrt{x}+\sqrt{y}}$$

thus

$$\sqrt{10002}-\sqrt{10001}=\frac{10002-10001}{\sqrt{10002}+\sqrt{10001}}=\frac{1}{\sqrt{10002}+\sqrt{10001}}$$

Now any loss in accuracy will just be round-off error. Here we get the value

$$D=4.999625078\times 10^{-3}$$

with all leading significant digits now being correct.

This idea of rationalization can be applied to the problem of solving quadratic equations.

Example 1.4: We recall the quadratic formula which states that the two solutions to the equation $ax^2+bx+c=0$, with $a\neq 0$, are given by

$$x=\frac{-b\pm\sqrt{b^2-4ac}}{2a}. \tag{1.3}$$

If $|4ac|$ is very small when compared to b^2, then the square root will look roughly like $|b|$. Thus, one of the roots of the equation will be approximately equal to $-\frac{b}{a}$.

As an example consider the equation

$$x^2-10^5x+1=0.$$

The computer or calculator described above, using (1.3), gives values for the roots

$$x_1=\frac{+10^5-\sqrt{10^{10}-4}}{2}\approx 0$$

and

$$x_2 = \frac{+10^5 + \sqrt{10^{10} - 4}}{2} \approx 10^5 .$$

This is because, compared to 10^{10}, 4 is very small and so we find that x_1 is approximately zero (or equal to zero if we don't keep enough places) and x_2 is approximately 10^5. If we go through the rationalization process to avoid the cancellation in the computation of x_1, we get

$$\begin{aligned} x_1 &= \frac{+10^5 - \sqrt{10^{10} - 4}}{2} = \frac{10^5 - \sqrt{10^{10} - 4}}{2} \cdot \frac{10^5 + \sqrt{10^{10} - 4}}{10^5 + \sqrt{10^{10} - 4}} \\ &= \frac{10^{10} - 10^{10} + 4}{2\left(10^5 + \sqrt{10^{10} - 4}\right)} = \frac{2}{10^5 + \sqrt{10^{10} - 4}} \approx 10^{-5} \end{aligned}$$

since even a rough approximation to the answer gives us x_1 approximately equal to 10^{-5}, which is much larger than 0. For the other root we can improve our accuracy by recalling that

$$ax^2 + bx + c = a(x - x_1)(x - x_2) = a\left(x^2 - (x_1 + x_2)x + x_1 x_2\right).$$

Thus,

$$x_1 + x_2 = -\frac{c}{a} \quad \text{and} \quad x_1 x_2 = \frac{c}{a} \tag{1.4}$$

so that

$$x_2 = +10^5 - x_1 \approx 10^5 - 10^{-5} .$$

We get the same result for x_2 using the second formula in (1.4).

Problem 1.2: Repeat the above argument for the equation $x^2 - 10^6 x + 1 = 0$ without using the formulas (1.5) and (1.6), below.

To summarize these results, if $b^2 - |4ac| \approx b^2$ we replace (1.3) by the new quadratic formula

$$x_1 = \begin{cases} \dfrac{2c}{-b + \sqrt{b^2 - 4ac}} & \text{if } b < 0 \text{ and} \\ \dfrac{2c}{-b - \sqrt{b^2 - 4ac}} & \text{if } b > 0 \end{cases} \tag{1.5}$$

$$x_2 = -\frac{b}{a} - x_1 \text{ or } x_2 = \frac{c}{ax_1}. \tag{1.6}$$

Formula (1.5) is obtained by rationalization, that is if $b < 0$,

$$x_1 = \frac{-b - \sqrt{b^2 - 4ac}}{2a} \left(\frac{-b + \sqrt{b^2 - 4ac}}{-b + \sqrt{b^2 - 4ac}} \right) = \frac{2c}{-b + \sqrt{b^2 - 4ac}}$$

and if $b > 0$,

$$x_2 = \frac{-b + \sqrt{b^2 - 4ac}}{2a} \left(\frac{-b - \sqrt{b^2 - 4ac}}{-b - \sqrt{b^2 - 4ac}} \right) = \frac{2c}{-b - \sqrt{b^2 - 4ac}}.$$

Thus, by this procedure, we are avoiding catastrophic cancellation by adding nearly equal numbers and not subtracting nearly equal numbers.

1.3 Ill-conditioned Problems

There are a variety of problems that are not "computationally continuous" with respect to the parameters of the problem. That is, a small change in the parameters of the problem lead to a large change in the output. As a simple example of this phenomenon we note that

$$P_1(x) = x^2 - 4x + 4 = 0$$

has a double root of $x = 2$. On the other hand, the polynomial

$$P_2(x) = P_1(x) - 10^{-8} = x^2 - 4x + 3.99999999 = 0$$

has roots $x_1 = 1.999683772$ and $x_2 = 2.000316228$ which differs significantly from the roots of P_1 in the sense that a change of 10^{-8} in one coefficient gives a change of more than 3×10^{-4} in the roots.

Note also, that if we take

$$P_3(x) = P_1(x) + 10^{-8}x = x^2 - 3.99999999x + 4\,,$$

then we would get no real roots since $(3.99999999)^2 - 4^2 < 0$. Figure 1.1, below, illustrates this state of affairs.

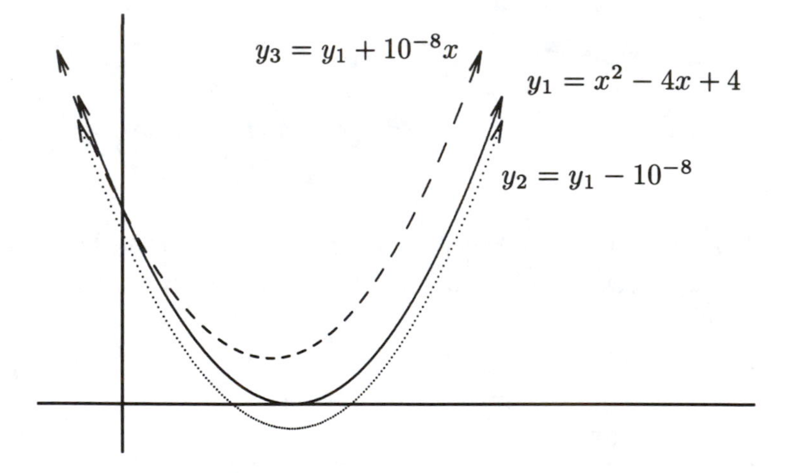

Figure 1.1

As another, more exotic example, we include the following celebrated example due to Wilkinson. This example shows that the instability can be even more widespread than just a change in the roots after a few decimal points. Let

$$P(x) = (x-1)(x-2)\cdots(x-19)(x-20) = x^{20} - 210x^{19} + \cdots .$$

It is easy to see that the roots of the polynomial are $x_1 = 1$, $x_2 = 2$, $\ldots$, $x_{19} = 19$ and $x_{20} = 20$. Suppose we now change the coefficient -210 to $-210 + 2^{-23} \approx -209.999999881$ and ask what happens to the roots. We find the following results:

$$\begin{aligned}
x_1 &= 1.000000000 & x_{11}, x_{12} &= 10.095266145 \pm 0.643500904i \\
x_2 &= 2.000000000 & x_{13}, x_{14} &= 11.793633881 \pm 1.652329728i \\
x_3 &= 3.000000000 & x_{15}, x_{16} &= 13.992358137 \pm 2.518830070i \\
x_4 &= 4.000000000 & x_{17}, x_{18} &= 16.730737466 \pm 2.812624894i \\
x_5 &= 4.999999928 & x_{19}, x_{20} &= 19.502439400 \pm 1.940330347i \\
x_6 &= 6.000006944 \\
x_7 &= 6.999697234 \\
x_8 &= 8.007267603 \\
x_9 &= 8.917250249 \\
x_{10} &= 20.846908101 .
\end{aligned}$$

Thus not only have some of the roots significantly changed in the first decimal place, but ten of the roots have become complex numbers, all of which are significantly off the real axis. The moral of the story is that while the roots of the equation do depend continuously on the coefficients, the dependence is quite sensative to small changes.

1.4 Numerical Instabilities

A major problem with computer mathematics is the accumulation of round-off error. By *round-off error* we mean the error in representing real numbers on a computer which has only a finite number of place values. For example, if we define

$$I_n = \int_0^1 \frac{x^n}{x+2}\, dx \tag{1.7}$$

for $n = 0, 1, 2, \ldots$, we note that $I_0 = \log \frac{3}{2}$ and that for $n \geq 1$ we have the following:

$$\begin{aligned} I_n &= \int_0^1 \frac{x^{n-1}(x+2-2)}{x+2}\, dx \\ &= \int_0^1 x^{n-1}\, dx - 2\int_0^1 \frac{x^{n-1}}{x+2}\, dx \\ &= \frac{1}{n} - 2I_{n-1}\,. \end{aligned} \tag{1.8}$$

Equation (1.8) is an example of a difference equation or recursive relation, and in theory, this equation and the value of I_0 are sufficient to determine the values of I_n for all $n = 0, 1, 2, \ldots$. Note that since $\frac{x^n}{x+2} \geq 0$ for $x \geq 0$ we must have

$$I_n \geq 0\,. \tag{1.9}$$

Also, since $x \geq 0$, we have $x + 2 \geq 2 > 0$ so that

$$\frac{x^n}{x+2} \leq x^n$$

for $x \geq 0$. Thus

$$I_n = \int_0^1 \frac{x^n}{x+2}\, dx \leq \int_0^1 x^n\, dx = \frac{1}{n+1}\,, \tag{1.10}$$

and since $x^{n+1} < x^n$ on $(0, 1)$ we see that

$$I_{n+1} = \int_0^1 \frac{x^{n+1}}{x+2} dx \leq \int_0^1 \frac{x^n}{x+2}\, dx = I_n\,.$$

Hence, as n tends to infinity I_n decreases monotonically to zero.

In fact in Exercise 1.5 we ask the reader to derive a better set of inequalities for I_n, namely, for $n = 0, 1, 2, \ldots$ we have

$$\frac{1}{3(n+1)} \le I_n \le \frac{1}{2(n+1)}. \tag{1.11}$$

We now use the recursive relation (1.8) to calculate the values of I_n for $n \ge 1$ and present an elementary error analysis for this example. To differentiate between the exact value I_n and the computed values, we will denote by J_n the computed values of I_n. For J_0 we take the computer value of $I_0 = \log \frac{3}{2}$. We then compute the values of J_n, $n \ge 1$, by the recursion formula (1.4). For $n = 0$ to 50, the values of J_n are given in the second column of Table 1.1.

We see from Table 1.1 that the values of J_n cease to be strictly positive for $n \ge 24$, in fact they alternate in sign. This violates the inequality (1.9). Also for $n \ge 24$ the values of J_n are seen to increase in absolute value and indeed for $n \ge 25$ the absolute value of J_n is no longer less than $\frac{1}{n+1}$, which violates the inequalities (1.11). One might reasonably ask at this point what went wrong. We shall do a simple error analysis to explain why this happened and then shall explain one way out of this difficulty.

Let $E_n = I_n - J_n$ for $n = 0, 1, 2, \ldots$ be the error at the nth step. To simplify our discussion, we shall assume that the only error in our calculations is due to the error $E_0 = I_0 - J_0$ and that the calculation of J_n, by the recursion formula

$$J_n = \frac{1}{n} - 2J_{n-1} \tag{1.12}$$

is exact for $n = 1, 2, \ldots$. In general, this is not quite true. There is usually an additional small error due to the operations in equation (1.12). With this simplifying assumption we have for $n = 1, 2, \ldots$

$$\begin{aligned} E_n = I_n - J_n &= \left(\frac{1}{n} - 2I_{n-1}\right) - \left(\frac{1}{n} - 2J_{n-1}\right) \\ &= -2(I_{n-1} - J_{n-1}) = (-2)E_{n-1}, \end{aligned} \tag{1.13a}$$

n	J_n	K_n	L_n	M_n
0	0.4054651084	0.4054651100	0.4054651100	0.4054651100
1	0.1890697900	0.1890697838	0.1890697838	0.1890697838
2	0.1218604200	0.1218604324	0.1218604324	0.1218604324
3	0.0896124933	0.0896124684	0.0896124684	0.0896124684
4	0.0707750134	0.0707750631	0.0707750631	0.0707750631
5	0.0584499732	0.0584498738	0.0584498738	0.0584498738
6	0.0497667202	0.0497669189	0.0497669189	0.0497669189
7	0.0433237024	0.0433233050	0.0433233050	0.0433233050
8	0.0383525952	0.0383533900	0.0383533900	0.0383533900
9	0.0344059207	0.0344043311	0.0344043311	0.0344043311
10	0.0311881586	0.0311913377	0.0311913377	0.0311913378
11	0.0285327737	0.0285264155	0.0285264155	0.0285264153
12	0.0262677859	0.0262805022	0.0262805023	0.0262805026
13	0.0243875051	0.0243620724	0.0243620723	0.0243620717
14	0.0226535612	0.0227044265	0.0227044268	0.0227044279
15	0.0213595442	0.0212578136	0.0212578130	0.0212578107
16	0.0197809116	0.0199843728	0.0199843740	0.0199843786
17	0.0192617062	0.0188547837	0.0188547814	0.0188547721
18	0.0170321431	0.0178459880	0.0178459926	0.0178460113
19	0.0185672927	0.0169396029	0.0169395936	0.0169395563
20	0.0128654146	0.0161207941	0.0161208128	0.0161208873
21	0.0218882184	0.0153774593	0.0153774220	0.0153772730
22	1.6781086E-03	0.0146996268	0.0146997013	0.0146999993
23	0.0401220436	0.0140790072	0.0140788581	0.0140782621
24	-0.0385774206	0.0135086522	0.0135089503	0.0135101423
25	0.1171548412	0.0129826955	0.0129820994	0.0129797153
26	-0.195848144	0.0124961474	0.0124973395	0.0125021078
27	0.428733325	0.0120447422	0.0120423580	0.0120328213
28	-0.8217523643	0.0116248012	0.0116295696	0.0116486431
29	1.6779874	0.0112331561	0.0112236194	0.0111854724
30	-3.32264161	0.010867021	0.0108860945	0.0109623885
⋮				
41	6827.033682	8.0118332E-03	-0.0310506666	-0.1873006665
42	-13654.0435	7.78585745E-03	0.0859108571	0.3984108569
43	27308.1102	7.684099E-03	-0.1485659004	-0.7735659
44	-5416.1977	7.35907475E-03	0.3198590736	1.56985907
45	109232.4176	7.50407275E-03	-0.617495925	-3.11749592
46	-218464.8135	6.73098495E-03	1.25673098	6.25673098
47	436929.6482	7.81462585E-03	-2.49218537	-12.49218537
48	-873859.2756	5.2040816E-03	5.00520408	25.00520408
49	1747718.57	0.01	-9.99	-49.99
50	-3495437.12	0	20	100

Table 1.1

so that $E_1 = -2E_0$, $E_2 = -2E_1 = 4E_0$ and, in general,

$$E_n = (-2)^n E_0 . \tag{1.13b}$$

Thus, any initial nonzero error E_0 leads quickly to a totally unacceptable answer. This explains, partly, why the values, once they start growing, grow so rapidly. Note that no matter how many significant places the computer carries, the error will still quickly get out of hand. Note that even if $E_0 = 0$, which is not possible in this case, we would soon pick up a small error which would then be magnified as above by the factor of -2. Indeed, the values for J_n in Table 1.1 indicate that for $n \geq 30$ the values are essentially E_n.

Problem 1.3: Use (1.11) to justify the claim in the last sentence for $n = 44$.

In many problems nothing can be done about the problem of instability. In our example problem we can calculate I_n accurately by rewriting (1.8) as

$$I_{n-1} = \frac{1}{2}\left(\frac{1}{n} - I_n\right) \text{ for } n \geq 1 . \tag{1.14}$$

We define the sequence $\{K_n\}_{n=0}^{50}$ by

$$K_{n-1} = \frac{1}{2}\left(\frac{1}{n} - K_n\right), \quad K_{50} = 0 . \tag{1.15}$$

These are the computer values of I_n and given in the third column of Table 1.1. The values are computed from the bottom to the top of the third column. We choose $K_{50} = 0$ because I_n is close to zero by our estimate in (1.6).

Our error analysis for (1.14) and (1.15) now follows as before. Let $F_n = I_n - K_n$. Then $F_{n-1} = -\frac{1}{2}F_n$ so that if $m \geq n$, then $F_n = \left(-\frac{1}{2}\right)^{m-n} F_m$. By (1.10), $\frac{1}{153} \leq I_{50} \leq \frac{1}{102}$ and hence

$\frac{1}{153} \le F_{10} \le \frac{2^{-40}}{102}$. Since $2^{40} \approx 10^{12}$, if our errors are only due to F_{50} and not in the calculations in (1.11), the value is correct to the accuracy of our computer.

To stress the imporance of our error analysis we have defined two additional sequences $\{L_n\}_{n=0}^{50}$ and $\{M_n\}_{n=0}^{50}$. They each satisfy (1.15) except that $L_{50} = 20$ and $M_{50} = 100$. The error analysis is as in the last paragraph except that $I_{50} - L_{50} \approx -20$ and $I_{50} - L_{50} \approx -100$. The values of the sequences are given in the last two columns in Table 1.1. As before, the calculations are from bottom to top. Note that regardless of the initial error, the factor of $\left(\frac{1}{2}\right)^n$ becomes so dominant that K_n, L_n and M_n agree for $n \le 9$ to the accuracy of the computer.

Problem 1.4: Carefully repeat, for the sequence L_n, the error analysis that was done for the sequence $\{F_n\}$ in the next to the last paragraph.

Exercise Set 1

1 Show that there are 49 possible numbers as described in the text using (1.1). What is the smallest and largest distance between these numbers?

2 Write and run a program that computes the two series

a) $1 + \frac{1}{2} + \frac{1}{3} + \frac{1}{4} + \frac{1}{5} + \frac{1}{6} + \cdots + \frac{1}{n} + \cdots$

and

b) $1 - \frac{1}{2} + \frac{1}{3} - \frac{1}{4} + \frac{1}{5} - \frac{1}{6} + \cdots + \frac{(-1)^{n+1}}{n} + \cdots$

to an accuracy of eight places. How many terms are needed in each case? (Note that the series in (a) diverges while the series in (b) converges. Nevertheless the computer will be able to find a sum for the series in (a).)

3 Using ten significant figures solve the equation $x^2 + 10^5 x + 1$ using Formula (1.3), then use Formulas (1.5) and (1.6). Explain the differences in the two methods.

4. Compute the solutions to $x^2 - 1634x + 2 = 0$. Use the usual quadratic formula (1.3) and then use a method to avoid catastrophic cancellation.

5. Show that if

$$I_n = \int_0^1 \frac{x^n}{x+2}\, dx\,,$$

then for $n = 0, 1, \ldots$ we have the inequalities

$$\frac{1}{3(n+1)} \leq I_n \leq \frac{1}{2(n+1)}\,.$$

6. Apply the binomial theorem to $(x + 2 - 2)^n$ to show that for $n = 1, 2, 3, \ldots$ we have

$$\int_0^1 \frac{x^n}{x+2}\, dx = \sum_{k=0}^{n-1} (-1)^k 2^k \binom{n}{k} \left(3^{n-k} - 2^{n-k}\right) \frac{1}{n-k} + (-2)^n \log\frac{3}{2}\,.$$

This gives us another expression for I_n which may be better for the purpose of calculation.

7. Let $I_n = \int_0^1 x^n e^x\, dx$.

   a) Show that $I_1 > I_2 > \cdots > 0$.

   b) Use integration by parts to show that $I_{n+1} = e - (n+1)I_n$ for $n \geq 1$. Show that $I_1 = 1$.

c) Using the recursion relation in part (b) calculate $I_1, I_2, \ldots, I_{49}, I_{50}$. Explain why they violate the inequalities of part (a).

d) Find a recusion relation that expresses I_{n-1} in terms of I_n.

e) Using a starting value of $I_{50} = 0$, use the recursion of part (d) to calculate $I_{49}, \ldots, I_2, I_1$. Explain your results.

f) Repeat part (e) with starting values of $I_{50} = 10$ and $I_{50} = 100$. Explain the results obtained. The results should be presented in a manner similar to Table 1.1.

8. The Fibonacci sequence is defined by $F_0 = F_1 = 1$ and for $n \geq 2$ by

$$F_n = F_{n-1} + F_{n-2} \tag{*}$$

It can also be shown that

$$F_n = \frac{1}{\sqrt{5}}\left[\left(\frac{1+\sqrt{5}}{2}\right)^n - \left(\frac{1-\sqrt{5}}{2}\right)^n\right]. \tag{**}$$

Use both relations (*) and (**) to calculate $F_0, F_1, \ldots, F_{50}$. Explain any difference in the results obtained. What conclusions can you draw about the closed form solution (**) as opposed to the recursion relation (*)? Contrast the effort needed for each relation. (HINT: to ease calculations note that $\frac{1-\sqrt{5}}{2} = \frac{-2}{1+\sqrt{5}}$.) Note that from (*) each F_n is an integer. However, if we compute F_n using (**) it is unlikely we will get an integer.

9. Compute the binomial coefficients $\binom{n}{m}$ by using the following three methods:

   a) Use the definition $\binom{n}{m} = \frac{n!}{m!(n-m)!}$. (Recall that $0! = 1! = 1$ and for $n \geq 2$, $n! = n(n-1)\cdots 2 \cdot 1$.)

   b) Show that $\binom{n}{m+1} = \frac{n-m}{m+1}\binom{n}{m}$. Use this result alone along with the fact that $\binom{n}{0} = 1$ for $n \geq 0$ to compute the binomial coefficients.

   c) Prove Pascal's relation $\binom{n}{m+1} = \binom{n-1}{m} + \binom{n-1}{m-1}$. Use this result alone along with the facts that $\binom{n}{0} = 1$ and $\binom{n}{m} = 0$ if $m > n$ to compute the binomial coefficients.

   If you calculate $\binom{n}{m}$ for $n = 1, 2, \ldots, 10$ and $m = 1, 2, \ldots, n$ which of these methods is the most accurate and which of them is the least? (Note that the identity in (c) is the basis of Pascal's triangle.)

10. Solve the system of equations

$$\begin{aligned} x + y &= 1 \\ x + (1-a)y &= 0\,, \end{aligned}$$

   for $a = 0.000001$ and for $a = 0.000011$. How close are the two sets of solutions?

11. Evaluate the infinite sum

$$\sum_{k=1}^{+\infty} \frac{1}{k(k+1)}$$

   to eight significant figures. How many terms will be needed? Would this problem have been easier to do if you had noticed that

$$\frac{1}{k(k+1)} = \frac{1}{k} - \frac{1}{k+1}\,?$$

12. Calculate the two sums, *in the order stated,*

$$T = \sum_{n=1}^{40} \frac{1}{n(n+1)(n+2)(n+3)}$$

and

$$U = \sum_{n=40}^{1} \frac{1}{n(n+1)(n+2)(n+3)} .$$

Did you find any differences in the values? If not try more precision.

2 *Review of Calculus*

The purpose of this brief chapter is to recall some important ideas from Calculus. In Section 2.1, we consider the Taylor series expansion for a function $f(x)$ defined on an interval of the real line. It allows us to approximate $f(x)$ by a polynomial. Although not usually stressed in Calculus, the remainder term $R_n(x, h)$ of the Taylor series expansion is important since it allows us to estimate the error in our approximation. Good numerical analysis requires that we not only make good approximations but (whenever possible) estimate our errors.

The second section is entitled: *Continuity Theorems.* Here we consider the Intermediate Value Theorem and various mean value theorems. They are primarily theorems of existence. These ideas occur repeatedly throughout all of advanced mathematics because most problems are impossible to solve exactly and do not yield closed form solutions. Before we can try to approximate the solutions we must be certain that these solutions exist. For example, it would be foolhardy to try to maximize the function $f(x) = x^2$ over all the real numbers.

2.1 Taylor Series Expansion

Perhaps the most important mathematical tool in numerical work is the Taylor series expansion which allows us to approximate a function by a polynomial with well determined coefficients.

The Taylor series expansion of a function $f(x)$ with step size h about x is

$$\begin{aligned} f(x+h) &= f(x) + hf'(x) + \frac{h^2}{2!}f''(x) + \cdots \\ &\quad + \frac{h^n}{n!}f^{(n)}(x) + R_n(x,h) \\ &= \sum_{k=0}^{n} \frac{h^k}{k!} f^{(k)}(x) + R_n(x,h) \end{aligned} \tag{2.1}$$

where we assume that the function f has continuous derivatives of order $n+1$ in a neighborhood of x and the expression $R_n(x,h)$ is the error term. It can be shown that

$$R_n(x,h) = \frac{h^{n+1}}{(n+1)!} f^{(n+1)}(\xi) \ ,$$

where ξ is between x and $x+h$. We note that ξ depends on h, x and n.

Of special note is that the difference

$$P_n(h) = f(x+h) - R_n(x,h)$$

is a polynomial of degree n in h. If $n = 1$, then $P_1(h) = f(x) + hf'(x)$ is the equation of the tangent line to the graph of f at the point x. For our purposes this line is the "best linear approximation" to the curve at that point. If $n = 2$, then $P_2(h) = f(x) + hf'(x) + \frac{h^2}{2}f''(x)$ is the parabola of best approximation to the curve at the point x.

Problem 2.1: Find the best approximation for $f(x) = 2x^2 - 3x + 4$, $x = 0$ and $n = 2$. Did you expect this result for this example?

If f has derivatives of all order, then we may formally replace (2.1) with

$$
\begin{aligned}
f(x+h) &= f(x) + hf'(x) + \cdots + \frac{h^n}{n!} f^{(n)}(x) + \cdots \\
&= \sum_{k=0}^{+\infty} \frac{h^k}{k!} f^{(k)}(x) .
\end{aligned} \tag{2.2}
$$

We remind the reader that infinite series such as (2.2) present special problems of convergence which are covered in the standard calculus courses, and so we will not deal with them in this text. The expression (2.1) is often written

$$
f(x) = \sum_{k=0}^{n} \frac{f^{(k)}(x_0)}{k!} (x - x_0)^k + R_n(x, x_0) \tag{2.3a}
$$

where we have replaced h by $x - x_0$ and x by x_0, to emphasize the idea that the point $x = x_0$ is fixed. Here we have

$$
R_n(x, x_0) = \frac{(x - x_0)^{n+1}}{(n+1)!} f^{(n+1)}(\xi) \tag{2.3b}
$$

where ξ is between x and x_0.

The most common form of (2.3) is the Maclaurin series,

$$
f(x) = \sum_{k=0}^{n} \frac{f^{(k)}(0)}{k!} x^k + R_n(x) , \tag{2.4}
$$

which is formed by setting $x_0 = 0$ in equation (2.3).

For example, to find an approximation to $\sqrt{10001}$ we may set $f(x) = (10000 + x)^{\frac{1}{2}}$ and obtain

$$f'(x) = \frac{1}{2}(10000 + x)^{-\frac{1}{2}}.$$

Thus $f(0) = 100$, $f'(0) = \frac{1}{200}$, and our linear approximation is, by (2.4),

$$(10000 + x)^{\frac{1}{2}} \approx 100 + \frac{x}{200} = 100 + .005x.$$

We note once again that $P_1(x) = 100 + .005x$ is the tangent line to $(10000 + x)^{\frac{1}{2}}$ at the point $x_0 = 0$. We are interested in the approximation when $x = 1$, which gives $\sqrt{10001} \approx 100.005$. The picture below, which is not true to scale, illustrates this state of affairs:

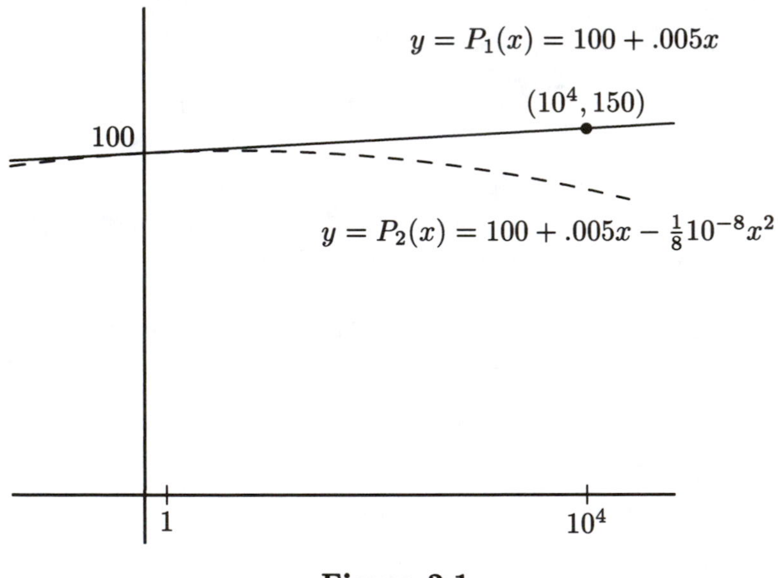

Figure 2.1

In Exercise 2.2 we ask the student to do this example again by using (2.3) with $f(x) = \sqrt{x}$ and $x_0 = 10000$. The corresponding picture will be the same as our picture except that the independent variable x has been increased, or translated, by 10000.

To try to get a better approximation of $\sqrt{10001}$ we take $n = 2$ in (2.3). We have $f(x) = (10000 + x)^{\frac{1}{2}}$, $f'(x) = \frac{1}{2}(10000 + x)^{-\frac{1}{2}}$ and $f''(x) = -\frac{1}{4}(10000 + x)^{-\frac{3}{2}}$. Thus $f(0) = 100$, $f'(0) = \frac{1}{200}$ and $f''(0) = -\frac{1}{4}10^{-6}$. Therefore, our parabolic approximation to $(10000 + x)^{\frac{1}{2}}$ is

$$
\begin{aligned}
P_2(x) &= f(0) + f'(0) + \frac{x^2}{2} f''(0) \\
&= 100 + .005x - (1.25 \times 10^{-7})x^2 .
\end{aligned}
$$

At $x = 1$ we see that $\sqrt{10001} \approx 100 + .005 - (1.25 \times 10^{-7}) = 100.004999875$. We invite the reader to use (2.3b) to show that the error is less than 5×10^{-10}. In the figure above we have sketched $P_2(x)$. The two polynomials are almost identical near $x = 0$ since $f(x)$ is almost linear close to $x = 0$. If x is far from zero, say $x = 10^4$, the two polynomials will give quite different values as our picture shows. Finally, we note that $P_1(x)$ is the tangent line of $P_2(x)$ and all higher order approximating polynomials $P_n(x)$ at $x = 0$.

As a second example, we consider the Maclaurin series of $f(x) = \sin x$. If we take $f(x) = \sin x$ we have the following table:

n	$f^{(n)}(x)$	$f^{(n)}(0)$	$\frac{f^{(n)}(0)}{n!}$
0	$\sin x$	0	0
1	$\cos x$	1	1
2	$-\sin x$	0	0
3	$-\cos x$	-1	$-\frac{1}{6}$

Table 2.1

Since $f^{(4)}(x) = \sin x$, we see that $f^{(n+4)}(0) = f^{(n)}(0)$ for $n = 0, 1, 2, \ldots$. Thus,

$$\sin x = x - \frac{x^3}{3!} + \frac{x^5}{5!} - \cdots = \sum_{k=0}^{+\infty} (-1)^k \frac{x^{2k+1}}{(2k+1)!},$$

at least in a formal sense. The ratio test shows that this infinite series converges for all values of x since

$$\frac{x^2}{(2k+2)(2k+3)} \to 0 \quad \text{as} \quad k \to \infty \quad \text{for all real } x.$$

Problem 2.2: Find the approximating polynomial of degree at most two for $f(x) = \sin x$ at $x = 0$.

Problem 2.3: Use the ratio test to show that the power series expansion for $\sin x$ at $x = 0$ converges for all x.

We note that if we had used the remainder term in (2.3b) we get the equality

$$R_n(x, 0) = \frac{x^{n+1}}{(n+1)!} \sin^{(n+1)}(\xi).$$

Since the derivatives of $\sin x$ are plus or minus $\sin x$ or $\cos x$ and therefore bounded in absolute value by one we have

$$|R_n(x, 0)| \leq \frac{x^{n+1}}{(n+1)!}$$

which goes to zero for all real x.

To help recall some of the other facts about infinite series we find it is instructive to use the series expansion above for $\sin x$ to give some inequalities such as $\sin x < x$ for $x > 0$. The basic result we will need is the following theorem which says, in words, that when dealing with alternating series the error made by stopping at the nth term is, in absolute value, at most the absolute value of the $(n+1)$st term.

Theorem 2.1: (Alternating series test). *Let* $a_0 > a_1 > a_2 > \cdots > 0$. *If* $\lim_{n\to\infty} a_n = 0$, *then* $S = \sum_{k=0}^{+\infty}(-1)^k a_k$ *converges. If* $S_n = \sum_{k=0}^{n}(-1)^k a_k$ *then we have the estimate* $|S - S_n| < a_{n+1}$.

To apply this to our problem first note that if $x \geq 1$, then $\sin x < x$ since $|\sin x| \leq 1$ for all x and $\sin 1 < 1$. For $0 < x < 1$, we first note that the series for $\sin x$ is alternating and that

$$-\frac{x^3}{3!} + \frac{x^5}{5!} < 0\,, \quad -\frac{x^7}{7!} + \frac{x^9}{9!} < 0\,, \quad \ldots$$

so that $\sin x < x$. The inequality $x - \frac{x^3}{3!} < \sin x$ on $(0,1)$ is obtained since

$$\frac{x^5}{5!} - \frac{x^7}{7!} > 0\,, \quad \frac{x^9}{9!} - \frac{x^11}{11!} > 0\,, \quad \ldots$$

on $(0,1)$.

In Exercise 2.2 we ask the reader to generalize these results. These results are used to decide how many terms we need to find $\sin x$ to a desired accuracy for x in $(0,1)$. Thus, for example, to find $\sin x$ good to 12 significant figures in $(0,1)$, by Theorem 2.1, we choose n large enough so that $\frac{1}{(2n+3)!} < \frac{1}{2}10^{-12}$.

Assuming we have a valid infinite series expansion for a function, we continue our discussion of how many terms of the Taylor series are needed to get a given error. We begin with a straightforward problem of finding n so that the approximation of $f(x) = e^x$ by $P_n(x) = 1 + x + \cdots + \frac{x^n}{n!}$ is good to 12 significant figures for $0 \leq x \leq 1$. Our expansion is no longer an alternating series so we will use the remainder formula of the Taylor series expansion to find the value of n so that

$$|R_n(x)| < \frac{1}{2}10^{-12}$$

for all x in the interval $[0,1]$. Since $f(x) = e^x$ satisfies $f^{(n)}(x) = e^x$ for all values of n, we require that

$$|R_n(x)| = \left|\frac{e^\xi}{(n+1)!}\right| < \frac{1}{2}10^{-12},$$

where ξ is a point in the interval $(0, x)$. Since e^x is positive and increasing on $[0,1]$ this requirement will be met by any n that satisfies

$$\frac{e}{(n+1)!} < \frac{1}{2}10^{-12}$$

or

$$(n+1)! > (2e)10^{12}.$$

Since our factorial tables show that $16! > 2.0922 \times 10^{13}$ we see that we can take $n = 15$. Thus, we can approximate e^x to 12 significant figures uniformly in x (that is, with no dependencies on x) on $0 \leq x \leq 1$ by the polynomial

$$P_{15}(x) = 1 + x + \frac{x^2}{2} + \cdots + \frac{x^{15}}{15!}.$$

A slightly more difficult problem is to determine the number of terms needed when this number depends on the dependent variable x. For example, if we expand $f(x) = \ln x$ about $x = 1$ we obtain Table 2.2, below. Thus, the Taylor series for $\ln x$ about $x = 1$ is

$$\log x = (x-1) - \frac{1}{2}(x-1)^2 + \frac{1}{3}(x-1)^3 - \cdots + \frac{(-1)^{k+1}}{k}(x-1)^k + \cdots.$$

We note in passing that by the ratio test the series converges for $|x-1| < 1$ or $0 < x < 2$. By Theorem 2.1, this series converges for $x = 2$. The series diverges for $x = 0$ as it becomes the negative of the harmonic series encountered in calculus.

n	$f^{(n)}(x)$	$f^{(n)}(1)$	$\frac{f^{(n)}(1)}{n!}$
0	$\ln x$	0	0
1	x^{-1}	1	1
2	$-x^{-2}$	-1	$-\frac{1}{2}$
3	$2x^{-3}$	2	$\frac{1}{3}$
$\vdots$	$\vdots$	$\vdots$	$\vdots$
k	$(-1)^{k+1}(k-1)!x^{-k}$	$(-1)^{k+1}(k-1)!$	$\frac{(-1)^{k+1}}{k}$

Table 2.2

As in the last examples, we wish to find a value of n so that the polynomial

$$P_n(x) = (x-1) - \frac{1}{2}(x-1)^2 + \cdots + \frac{(-1)^{n+1}}{n}(x-1)^n \tag{2.5}$$

gives 12 place accuracy for the function $\log x$ on the interval $0 < x \leq 2$. Proceeding as above, we have

$$|R_n(x)| = \left| \frac{(-1)^{n+2} n! \xi^{-n-1}}{(n+1)!}(x-1)^{n+1} \right|$$

where ξ is between 1 and x, and so we need to find n so that

$$\left| \frac{(x-1)^{n+1}}{n\xi^{n+1}} \right| \leq \frac{1}{2} 10^{-12}.$$

Now, since $0 < x \leq 2$ we have that

$$\left| \frac{(x-1)^{n+1}}{n\xi^{n+1}} \right| \leq \frac{1}{n\xi^{n+1}}.$$

If $1 \leq x \leq 2$ then our last inequality is bounded by $\frac{1}{n}$ and as before we may choose $n > 2 \cdot 10^{12}$. If $0 < x < 1$ there are special problems, since if x approaches 0, ξ may approach 0 and $\frac{1}{\xi^{n+1}}$ is unbounded. Thus, a value of n can not be obtained for all x on $(0, 1)$. In order to find an n we must restrict ourselves to an interval $[\delta, 1)$ where $0 < \delta < 1$. In this case we choose n so that

$$|R_n(x)| \leq \frac{1}{n\xi^{n+1}} \leq \frac{1}{n\delta^{n+1}} \leq \frac{1}{2}10^{-12}.$$

Problem 2.4: Repeat all the results of the last two paragraphs for $\ln x$ about $x = 2$.

2.2 Continuity Theorems

There are several important results for continuous functions which are necessary for this course.

Theorem 2.2: *If $f(x)$ is continuous on the interval $[a, b]$, then $f(x)$ assumes its maximum and minimum values on $[a, b]$.*

Note that Theorem 2.2 is an existence theorem, that is it guarantees that there exist points t_1 and t_2 in $[a, b]$ such that

$$f(t_1) \leq f(x) \leq f(t_2)$$

for all x in $[a, b]$. However, the theorem tells you nothing about how to find the values of t_1 and t_2 nor about the uniqueness of t_1 and t_2.

If we consider $f(x) = x^2$ on the open interval $(0, 1)$, then $f(x)$ does not take on a maximum nor minimum value on $(0, 1)$, but does obtain these values on the closed interval $[0, 1]$. Likewise, if $g(x) = x^2$ on $[0, 1)$ and $g(1) = 0$ then $g(x)$ has no maximum. The problem is, of course, that $g(x)$ is not continuous on $[0, 1]$.

Theorem 2.3: (Intermediate Value Theorem). *If $f(x)$ is continuous on $[a,b]$ and $a \leq t_1 < t_2 \leq b$, then f assumes all values between $f(t_1)$ and $f(t_2)$.*

If, for example, $f(t_1) \leq f(t_2)$, then for any value y such that $f(t_1) \leq y \leq f(t_2)$, then there exists a t in the interval (t_1, t_2) such that $f(t) = y$. This theorem is particularly useful in finding roots of continuous functions, where $f(t_1)$ and $f(t_2)$ are of opposite signs. This leads to the method of bisections which will be discussed in Chapter 3.

Note that the theorem does not say there is a unique value of t that gives the value y.

Theorem 2.4: (Mean Value Theorem). *If f is continuous on $[a,b]$ and differentiable on (a,b), then there exists an ξ in (a,b) such that*

$$\frac{f(b)-f(a)}{b-a} = f'(\xi)\,.$$

This theorem has many applications in numerical analysis. Note that it follows from the case $n = 0$ of the Taylor series expansion, that is,

$$f(b) = f(a) + (b-a)f'(\xi)\,.$$

We can use this theorem to again prove that $\sin x < x$ for $x > 0$. For we have

$$\frac{\sin x - \sin 0}{x - 0} = \cos\xi\,.$$

Since $\cos\xi \leq 1$ at $x > 0$ we have

$$\frac{\sin x}{x} \leq 1$$

or $\sin x \leq x$.

The student will recall from calculus that a special case of the Mean Value Theorem is Rolle's Theorem. In addition to the hypothesis in Theorem 2.3 we must assume $f(a) = f(b)$. The conclusion is that $f'(\xi) = 0$ for some ξ, $a < \xi < b$.

The last two theorems are very useful in establishing at least one solution or at most one solution for an equation $f(x) = 0$. For example, if $f(x) = x^7 + 5x^3 + x - 6$ then $f(0) = -6 < 0$ and $f(1) = 1 > 0$ so that by the Intermediate Value Theorem there exists ξ, $0 < \xi < 1$ such that $f(\xi) = 0$. If $f(x_1) = f(x_2) = 0$ for $x_1 \neq x_2$ then by the Mean Value Theorem there exists ξ_1 such that

$$f'(\xi_1) = \frac{f(x_2) - f(x_1)}{x_2 - x_1} = 0$$

but $f'(x) = 7x^6 + 15x^2 + 1 > 0$ so that $f(x) = 0$ at most once. Thus, $f(x) = x^7 + 5x^3 + x - 6$ has exactly one root.

Problem 2.5: Suppose $P(x)$ is a polynomial with 4 distinct roots in an interval. State and justify a result for the number of roots of $P''(x)$ in the interval. Would the same result hold when $P(x)$ is replaced by a nonpolynomial function $f(x)$?

Problem 2.6: Repeat Problem 2.5 but assume that the roots were not distinct and occured at the point $\overline{x}$.

For completeness we include several theorems involving the integral of $f(x)$. For convenience, we assume that $f(x)$ and $g(x)$ are continuous on the interval $[a, b]$ when they are used in these equations. Thus, we will avoid questions of the existence of the definite integrals.

Theorem 2.5: *If* $|f(x)| \leq M$ *on* $[a, b]$ *for a constant* $M > 0$ *then*

$$\left| \int_a^b f(x)\, dx \right| \leq \int_a^b |f(x)|\, dx \leq M(b - a)\,.$$

Theorem 2.6: *There exist ξ in (a,b) such that*

$$\int_a^b f(x)\,dx = (b-a)f(\xi)\,.$$

Theorem 2.7: *If $m \le f(x) \le M$ on $[a,b]$ and $g(x) \ge 0$ on $[a,b]$ then*

$$m\int_a^b g(x)\,dx \le \int_a^b f(x)g(x)\,dx \le M\int_a^b g(x)\,dx\,.$$

Theorem 2.8: *If $g(x)$ does not change sign in $[a,b]$, then there exists ξ in (a,b) such that*

$$\int_a^b f(x)g(x)\,dx = f(\xi)\int_a^b g(x)\,dx\,.$$

Theorem 2.5 is immediate from Theorem 2.7 if $g(x) \equiv 1$. A similar comment holds for Theorem 2.6. The results are listed in the order above because of the special importance of Theorems 2.5 and 2.6 and the fact that their justification is easier than those of Theorems 2.7 and 2.8.

Problem 2.7: Show that Theorems 2.5 and 2.6 are corollaries (follow from) Theorems 2.7 and 2.8.

Exercise Set 2

1. Show that $\log 2 = 1 - \frac{1}{2} + \frac{1}{3} - \frac{1}{4} + \cdots$.

2. Find an approximation to $\sqrt{10001}$ by using equation (2.3) with $x_0 = 10000$ for the cases $n = 1$ and $n = 2$. Give an estimate of the error in each case. Sketch the corresponding picture for $n = 1$ and $n = 2$. (Hint: use the function $f(x) = \sqrt{x}$.)

3. Show that if $0 < x < 1$, then

$$x - \frac{x^3}{3!} < x - \frac{x^3}{3!} + \frac{x^5}{5!} - \frac{x^7}{7!} < \sin x < x - \frac{x^3}{3!} + \frac{x^5}{5!} < x\,.$$

   Generalize the above inequality to find better bounds for $\sin x$.

4. If $f(x) = 2x^3 - x^2 + 5x - 7$, find the Taylor series expansion of $f(x)$ about $x = 0$. Let $g(x)$ be the Taylor series expansion of $f(x)$ about $x = 1$. Find $f(2)$ and $g(2)$. Explain.

5. Let $f(x) = \cos x$. Find the Taylor series expansion about $x = 0$. How many terms of this series are required to ensure an accuracy of 12 significant digits if x is in the interval $\frac{1}{3} < x < \frac{1}{2}$?

6. In the previous chapter we introduced the integrals

$$I_n = \int_0^1 \frac{x^n}{x+2}\, dx$$

   for $n = 0, 1, 2, \ldots$.

   a) Recall from calculus the following theorem.

Theorem: *If the power series* $\sum_{k=0}^{+\infty} a_k(x-x_0)^k$ *converges for* $|x-x_0|<r$*, then, for* $|x-x_0|<r$*, we have*

$$\int \sum_{k=0}^{+\infty} a_k(x-x_0)^k \, dx = \sum_{k=0}^{+\infty} \frac{a_k}{k+1}(x-x_0)^{k+1}.$$

The theorem tells us we can integrate term by term as long as we stay within the interval of convergence. Use the fact that

$$\frac{1}{x+2} = \frac{1}{2} \cdot \frac{1}{1+\frac{x}{2}} = \frac{1}{2}\sum_{k=0}^{+\infty} \frac{(-1)^k x^k}{2^k}$$

for $|x| < 2$ to obtain an infinite series expansion for the integrals I_n as defined above. The series should not have any variables in it, that is, it should be a strictly numerical series involving n.

b) How many terms of the series obtained in (a) for I_n must be used to guarantee an accuracy of 12 significant digits for all values of n?

7. The Bessel function arises, for example, in the problem of wave propogation in three dimensions. If m is a nonnegative real number, then the Bessel function of order m can be shown to have the series expansion

$$J_m(t) = \left(\frac{t}{2}\right)^m \sum_{k=0}^{+\infty} (-1)^k \frac{(t/2)^{2k}}{k!(m+k)!}.$$

If we use the ratio test, we see that this series converges for all values of t for any value of $m \geq 0$. If we restrict the variable t to lie in the interval $[-1,1]$, how many terms of the series are needed to guarantee an accuracy of 12 significant digits for all values of t in $[-1,1]$ and all values of $m \geq 0$? Suppose we let t lie in the interval $[-1,2]$, what is the number of terms required in this case?

8. Prove Theorem 2.5 using Theorem 2.7.

9. Use Theorem 2.1 to ensure an accuracy of 12 significant digits if $P_n(x)$ as given in (2.5) is used to approximate $f(x) = \log x$ on the interval $[1, 2]$. Can you use this theorem on the interval $(0, 1)$?

10. Use the Mean Value Theorem or Rolle's Theorem to show that $f(x) = x^3 - x$ has a derivative $f'(x)$ which vanishes on $(-1, 0)$ and $(0, 1)$ and that $f''(x)$ vanishes on $(-1, 1)$.

11. Use the Mean Value Theorem (Theorem 2.4) to prove Rolle's Theorem: if f is continuous on $[a, b]$ and differentiable on (a, b) and $f(a) = f(b)$, then there is a point c in (a, b) such that $f'(c) = 0$.

12. Generalize Exercise 2.11 to the case when $f(x_i) = 0$ for $i = 1, \ldots, n$ and $x_1 < x_2 < \cdots < x_n$. What continuity conditions must f and its derivatives satisfy?

13. Use Rolle's Theorem to show that $f(x) = x^5 + 2x^3 - 2$ has exactly one root in the interval $[0, 1]$. Show further that this root is the only real root of $f(x)$.

14. Use Theorem 2.5 to show that

$$\left| \int_0^3 e^x \sin(x^2 \cos e^x) \, dx \right| \leq 3e^3 .$$

15. Verify the results 1–15 as stated in Section 2.3.

3 *Nonlinear Equations*

The purpose of this chapter is to consider how to solve nonlinear equations in one independent variable, such as

$$f(x) = x - \frac{3}{4}\sin x - \frac{2}{3} = 0\,. \tag{3.1}$$

Problems of this type are very important and occur in many areas of applied mathematics. In practice what often happens is that the person looking for a solution tries something like Newton's method. If it works, as it sometimes does, one congratulates oneself and goes on to another task, ignoring the possibility that there may be other solutions. If it doesn't work, one rechecks the program code (sometimes) and runs to find an expert in numerical analysis to find out what is wrong.

What should be done is to try to first understand the local behavior of $f(x)$ near the solutions. A sketch of $f(x)$ is best, but often not available, so we might restrict ourselves to an interval of interest such as $[-\pi, \pi]$. Several questions should be answered before any computer work is begun. Are there any solutions to (3.1)? If so, how many solutions are there on an interval such as $[-\pi, \pi]$? Is there a unique solution on this interval?

We will soon see that Equation (3.1) is solvable and in fact has a unique solution on $[-\pi, \pi]$. Now the real work begins. What method should we use? Is there a "best" method or at least a method which is easy to use and leads quickly to an approximate solution which is very close to the real solution, which we denote by r. We will soon see that the choice of a method to find a solution r such that $f(r) = 0$ involves the properties of f and the question of how close we must be to the solution r for our method to work? Will our method work if we start searching too far from r? How many steps or iterations do we need to get an acceptable error or are we likely to get into an infinite loop?

It is particularly surprising that while we need some mathematical theory, the level is often elementary and already available to the reader. In the above example, $f(-\pi) < 0$ and $f(\pi) > 0$, so that the Intermediate Value Theorem guarantees at least one solution on $[-\pi, \pi]$. On the other hand, $f'(x) = 1 - \frac{3}{4}\cos x > 0$ so there can be at most one real solution since the function f is strictly increasing. Thus, Equation (3.1) has exactly one solution on $(-\pi, \pi)$.

The topics in this chapter are as follows. In Section 3.1 we will use the Method of Bisection which is based on the Intermediate Value Theorem to find solutions of equations such as the equation in (3.1). In the remainder of this chapter we will derive more general ideas about iteration methods. These ideas will help us decide when equations such as (3.1) can be solved and how to solve them quickly and efficiently. In Section 3.2 we change the equation $f(x) = 0$ into an equivalent form $x = g(x)$ in order to obtain an iteration method where a sequence $\{x_n\}$ is determined by $x_{n+1} = g(x_n)$. Using the Mean Value Theorem we will see that good methods satisfy the condition $|g'(x)| \leq k < 1$ for x close to a solution r and that the condition $|g'(x)| \geq k > 1$ for x is close to a solution r indicates that the method is bad. In Section 3.3 we will consider the best known iteration method which is called Newton's Method or sometimes the Newton-Raphson Method. In Section 3.4 we will consider a method related to Newton's Method called the Secant Method. Section 3.5 contains the theoretical ideas for the earlier sections. These ideas have already been presented in a less formal manner to blend in with the ideas and examples of the earlier sections. Finally in Section 3.6 we consider the special problem of finding roots of a polynomial.

3.1 The Method of Bisection

The simplest method of finding an approximate solution (or root) of an equation $f(x) = 0$ is the bisection method. We require f to be continuous on $[a, b]$ and $f(a)$ and $f(b)$ to be opposite sign. The method is guaranteed to work, i.e. to find a value of r such that $f(r) = 0$, because of the Intermediate Value Theorem, although convergence is often very slow in comparison to other methods that we will consider.

Let us now see how easy it is to solve (0.1) using the Method of Bisection. We note that $f(x) = x - \frac{3}{4}\sin x - \frac{2}{3}$ is continuous on the interval $[-\pi, \pi]$ and that $f(-\pi) < 0$ while $f(\pi) > 0$. By the Intermediate Value Theorem, there exists at least one point r in $(-\pi, \pi)$ such that $f(r) = 0$. Setting $a_1 = -\pi$ and $b_1 = \pi$ we define $c_1 = \frac{a_1+b_1}{2}$. In this case $c_1 = 0$. Since $f(c_1) = -\frac{2}{3} < 0$ we know there is a root between $c_1 = 0$ and $b_1 = \pi$. Defining a_2 and b_2 by $a_2 = c_1 = 0$ and $b_2 = b_1$, our root x_0 is on the interval $[a_2, b_2]$. Setting $c_2 = \frac{a_2+b_2}{2} = \frac{\pi}{2}$ we have $f(c_2) = \frac{\pi}{2} - \frac{3}{4} - \frac{2}{3} > 0$, so our root is on the interval $[0, \pi/2]$. Defining $a_3 = a_2 = 0$ and $b_3 = c_2 = \frac{\pi}{2}$, we set $c_3 = \frac{a_3+b_3}{2}$.

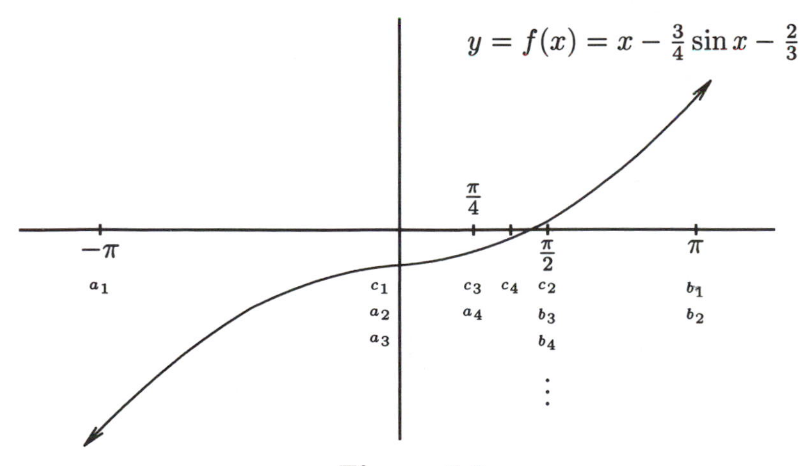

Figure 3.1

We continue this process with results listed in Table 3.1 below. At each step we generate three numbers a_n, b_n and c_n with $a_n < b_n$, $c_n = \frac{a_n+b_n}{2}$ and $b_n - a_n = \frac{b-a}{2^{n-1}} = \frac{\pi}{2^{n-2}}$. If $f(c_n) = 0$ for some n, we have found the root r. Otherwise, we may specify a value $\epsilon > 0$ so that our iteration terminates if $|a_n - b_n| < \epsilon$. Thus, the inequality $\frac{\pi}{2^{n-2}} < \epsilon$ implies that our iteration process terminates for the smallest natural number n which satisfies $n > 2 + \log_2 \frac{\pi}{\epsilon}$. As a practical matter we might specify a natural number N or a positive number $\epsilon_1 > 0$ so that our process also ends if $n > N$ or if $|f(c_n)| < \epsilon_1$. In Table 3.1 the first column lists the values of n, the second, third and fourth columns list the values of a_n, b_n, and c_n. The last column lists the values of $f(c_n)$. Thus, our approximation to r is $c_{28} = 1.40657630$ which satisfies $f(c_{28}) = 3.4 \times 10^{-9}$.

n	a_n	b_n	c_n	$f(c_n)$
1	-3.14159267	3.14159267	0.	-0.6666666666
2	0.	3.14159267	1.57079633	0.1541296634
3	0.	1.57079633	0.785398165	-0.4115985898
4	0.785398165	1.57079633	1.17809724	-0.1814790766
5	1.17809724	1.57079633	1.37444678	-0.0278088466
6	1.37444678	1.57079633	1.47262155	0.0595663434
7	1.37444678	1.47262155	1.42353416	0.0149851134
⋮				
16	1.40646862	1.40666037	1.40656449	-1.03566E-05
17	1.40656449	1.40666037	1.40661243	3.17034E-05
⋮				
27	1.40657628	1.40657637	1.40657632	2.34E-08
28	1.40657628	1.40657632	1.4065763	3.4E-09

Table 3.1

Problem 3.1: Use the values in Table 3.1 to determine a_6, a_7, b_6, b_7, c_6 and c_7 for the problem $g(x) = -f(x) = -x + \frac{3}{4}\sin x - \frac{2}{3}$.

In the general case where $f(x)$ is continuous on $[a, b]$ and $f(a)f(b) < 0$, there may be more than one root. Our iteration will converge to one of the roots on $[a, b]$. The procedure generates two sequences $\{a_n\}$ and $\{b_n\}$ with $a_1 = a$ and $b_1 = b$. At each step we define $c_n = \frac{a_n + b_n}{2}$. Our procedure terminates if $f(c_n) = 0$, $b_n - a_n < \epsilon$ where ϵ is a preassigned small positive number or if $n > N_{\max}$ as described above. See Exercise 3.3. Otherwise, we define

$$\begin{cases} a_{n+1} = c_n, \; b_{n+1} = b_n & \text{if} \quad f(c_n)f(b_n) < 0 \\ a_{n+1} = a_n, \; b_{n+1} = c_n & \text{if} \quad f(c_n)f(b_n) > 0 \,. \end{cases} \tag{3.2}$$

In closing this section, we note that the bisection method is preferred over iteration methods if we have little information about $f(x)$. The main disadvantages of this method are that it converges very slowly and the associated computer program is usually more complicated to write in that it may require some more logical steps.

3.2 The Method of Iteration

The purpose of this section is to consider iteration methods to solve the equation $f(x) = 0$. This is done by considering an equivalent equation $x = g(x)$ and then generating a sequence of real numbers $\{x_n\}$ where $x_n = g(x_{n-1})$ so that $x_n \to r$ where r is such that $f(r) = 0$. We will see that our major concern is whether g satisfies the condition $|g'(x)| \leq k < 1$ close to r for some positive number k. The formal theory is postponed until Section 3.5 so that the student can concentrate upon the examples and heuristic arguments.

If the equation $f(x) = 0$ is rearranged into the form of an equivalent equation

$$x = g(x)\,, \tag{3.3}$$

the equation (3.3) is then called *an iteration equation* for $f(x) = 0$. In particular, we note that $f(r) = 0$ if and only if $r = g(r)$. If $r = g(r)$ we say that r is a *fixed point* of $g(x)$. The *iteration method* is to find a function $g(x)$ and a starting value x_1 so that the sequence $\{x_n\}$ generated by $x_{n+1} = g(x_n)$, $n = 1, 2, 3, \ldots$ converges to r where $r = g(r)$ or equivalently $f(r) = 0$. This method is extremely powerful, but care must usually be taken to start close to the root. Thus, the Method of Bisection is often used in conjunction with the Method of Iteration to get close to the root.

Many examples are given below where we must start close to a root r, but sometimes for a fixed equation $f(x) = 0$ we can choose an iteration function $g_1(x)$ which works regardless of the initial value x_1. If, for example, our equation is $f(x) = x - \frac{3}{4}\sin x - \frac{2}{3} = 0$ with root r we claim that the iteration equation $x = g_1(x) = \frac{3}{4}\sin x + \frac{2}{3}$ generates a sequence $\{x_n\}$ so that $x_n \to r$ for any choice of the starting value x_1. This is illustrated in the second column of Table 3.2 where we start with $x_1 = 1000$. The reason that $g_1(x)$ is such a good iteration function is that $g_1'(x) = \frac{3}{4}\cos x$ or $|g'(x)| \leq \frac{3}{4}$. Thus, by the Mean Value Theorem if $|x_n - r|$ is the distance between the nth iteration value x_n and the root r we have

$$|x_{n+1} - r| = |g(x_n) - g(r)| = |g'(\xi)|\,|x_n - r| \leq \frac{3}{4}|x_n - r|$$

so that each succeeding distance is at most $\frac{3}{4}$ of the previous distance. By the completeness property of the real numbers we have $x_n \to r$. The reader should note that the final value in column two of Table 3.2 is the same as c_{28} in Table 3.1.

In the general case the question we must answer is what local properties of the function $g(x)$ near $x = r$ do we need to get a sequence $\{x_n\}$ which converges to a root r of $f(x)$? In Section 3.5 we will see that the sequence $\{x_n\}$ generated by $g(x)$ converges to a solution of $f(r) = 0$ if x_1 is in an interval I, centered at r and which for all x in I,

$$|g'(x)| \leq k < 1 \text{ for some constant } k. \tag{3.4}$$

n	$g_1 = \frac{3}{4}\sin x + \frac{2}{3}$	$g_2 = 2x - \frac{3}{4}\sin x - \frac{2}{3}$
1	1000	1.40657632
2	1.28677837	1.40657635
3	1.38661964	1.40657641
4	1.40398219	1.40657652
5	1.40625573	1.40657673
6	1.40653695	1.40657712
7	1.40657147	1.40657785
8	1.4065757	1.40657922
9	1.40657622	1.40658179
10	1.40657628	1.40658662
11	1.40657629	1.40659569
12	1.40657629	1.40661272
13	1.40657629	1.40664469
14	1.40657629	1.40670471
15	1.40657629	1.4068174
16	1.40657629	1.40702898
17	1.40657629	1.40742625
18	1.40657629	1.40817226
19	1.40657629	1.40957349
20	1.40657629	1.41220653
21	1.40657629	1.41715817
22	1.40657629	1.42648404
23	1.40657629	1.44409764
24	1.40657629	1.47754028
25	1.40657629	1.54167279
26	1.40657629	1.66699697
27	1.40657629	1.92079504
28	1.40657629	2.47039351

Note: The underlined digits are those that are in error.

Table 3.2

As in our example this result follows from the Mean Value Theorem since

$$\begin{aligned} |x_{n+1} - r| &= |g(x_n) - g(r)| \\ &= |g'(\xi)|\,|x_n - r| \\ &\le k|x_n - r| \\ &< |x_n - r|\,. \end{aligned}$$

Thus, each new iteration brings us closer to r. Repeated application of this inequality yields $|x_{n+1}-r| \le k|x_n - r| \le k^2|x_{n-1}-r| \le \dots \le k^n|x_1 - r|$, so that $|x_{n+1} - r| \to 0$.

Conversely, we may use the Mean Value Theorem to demonstrate that if

$$|g'(x)| \ge k > 1 \text{ for some constant } k, \tag{3.5}$$

the sequence $\{x_n\}$ cannot converge unless we choose $x_1 = r$, in which case $x_n = r$ for all $n = 1, 2, 3, \dots$. This is due to the fact that $|x_{n+1} - r| \ge k|x_n - r|$ and so even if we are close to r our distance from the root r increases with each iteration.

To illustrate this phenomena we choose $g_2(x) = 2x - \frac{3}{4}\sin x - \frac{2}{3}$ to be the iteration function for $f(x) = x - \frac{3}{4}\sin x - \frac{2}{3}$. In this case $g_2'(x) = 2 - \frac{3}{4}\cos x$ so that $g_2'(x) \ge \frac{5}{4}$ and hence if $x_{n+1} = g_2(x_n)$ then $|x_{n+1} - r| \ge \frac{5}{4}|x_n - r|$ and the iteration cannot converge to r if $x_1 \ne r$. This phenomena is exhibited in the third column of Table 3.2 where even though we set $x_1 = 1.40657632$, so that x_1 and r agree to seven figures, the sequence $\{x_n\}$ satisfies $|x_n - r| \to \infty$.

Several important ideas are illustrated by the next figure. We note that the "y-value" corresponding to an x-value of x_n is $y = x_n$ using the $y = x$ curve and $y = x_{n+1}$ using the $y = g(x)$ curve. Let $g(x)$, r_1 and r_2 be as in this figure and satisfy $x = g(x)$, $|g'(x)| \le k_1 < 1$ in (a, b) and $|g'(x)| \ge k_2 > 1$ in (c, d).

In our picture, $-1 < k_1 \le g'(x) < 0$ in (a, b) so that $|x_{n+1} - r| \le k_1|x_n - r|$ and the sequence $\{x_n\}$ converges to r_1, alternating around r_1. In the interval (c, d), $g'(x) \ge k_2 > 1$ so that no matter how close

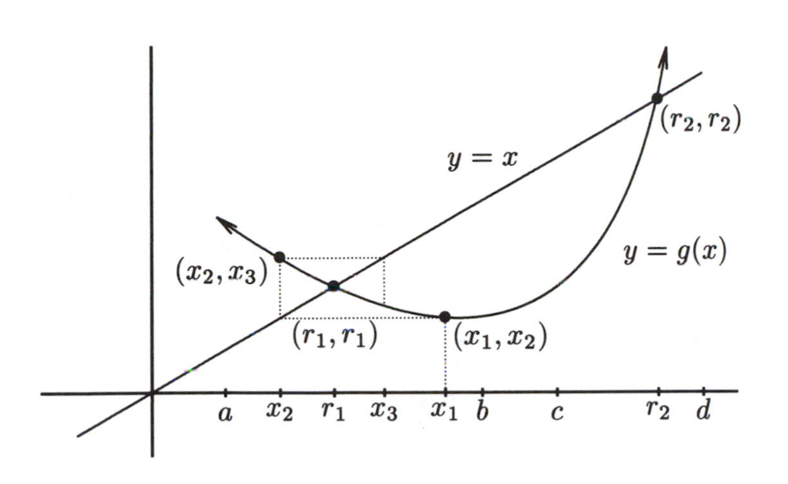

Figure 3.2

we start to r_2, if $x_1 \neq r_2$ the sequence $\{x_n\}$ can never converge to r_2. It is possible that we could start close to r_2 and converge to r_1 if some x_n is in (a, b).

For completeness we now picture four separate cases. In Case 1 we assume $g'(x) \geq k > 1$. As the reader can see $x_{n+1} - r > x_n - r$ so that $x_n \to \infty$ if $x_1 > r$ and $x_n \to -\infty$ if $x_1 < r$. In Case 2 we assume $0 < g'(x) < k < 1$ so that $x_n \to r$ with $r < x_{n+1} < x_n$ if $r < x_1$ or $x_n < x_{n+1} < r$ if $x_1 < r$. In Case 3 we assume $-1 < k \leq g'(x) < 0$ so that $x_n \to r$, alternating around r. In Case 4 we assume $g(x) \leq k < -1$ so that $|x_n| \to \infty$, alternating around r. That is, $x_{2n} \to \infty$ and $x_{2n+1} \to -\infty$ or $x_{2n} \to -\infty$ and $x_{2n+1} \to \infty$.

Problem 3.2: Sketch the appropriate curves in the four cases associated with Figure 3.3, below, for $h(x) = -g(x)$.

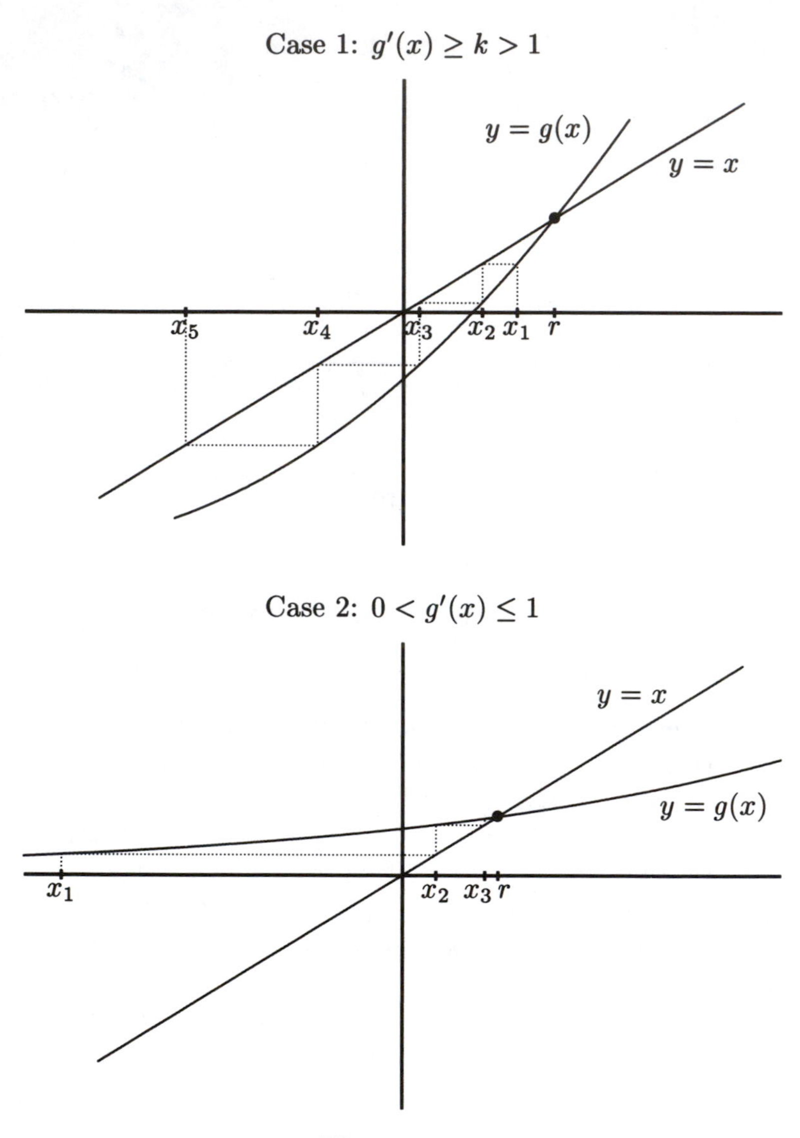
Case 1: $g'(x) \geq k > 1$
$y = g(x)$
$y = x$
x_5
x_4
x_3
x_2
x_1
r
Case 2: $0 < g'(x) \leq 1$
$y = x$
$y = g(x)$
x_1
x_2
x_3
r

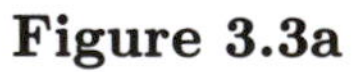

Figure 3.3a

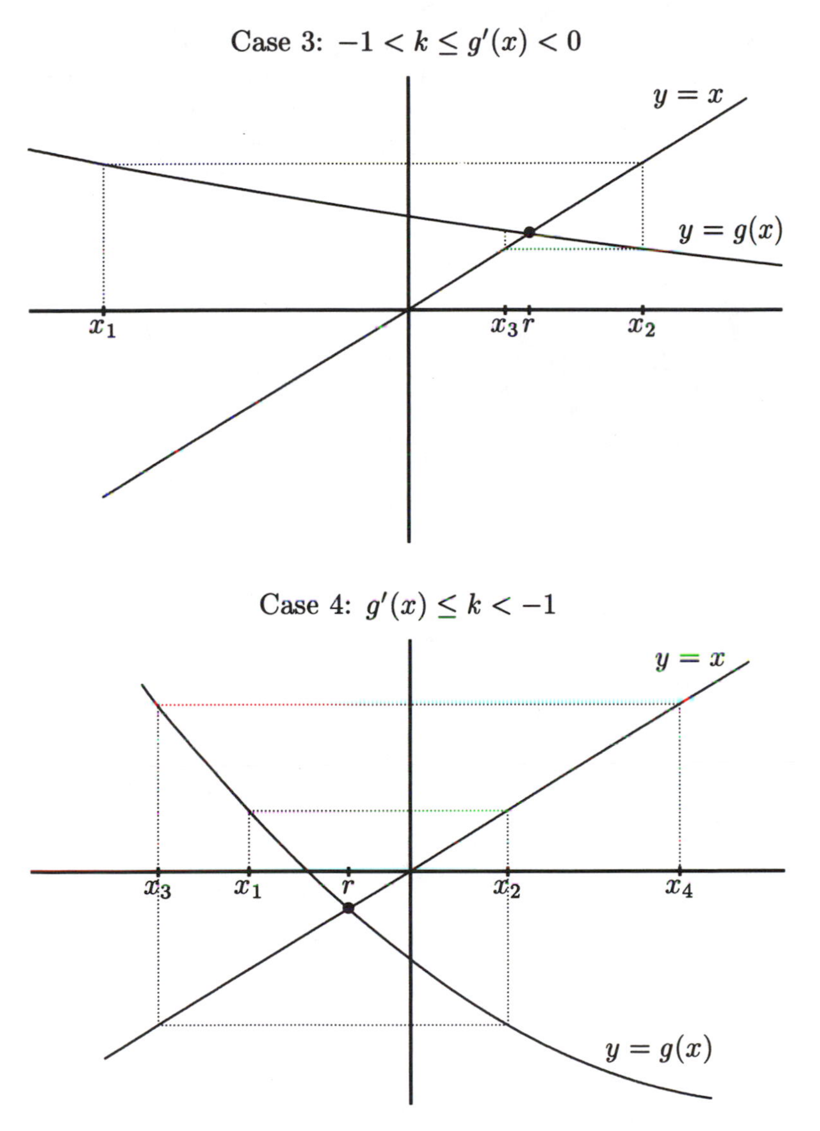

Figure 3.3b

The necessity of starting close to the root of $f(x)$ is very important as we shall see in the following example which has several parts, some of which we ask the reader to work. Let $f(x) = x^2 - x - \frac{15}{4}$ which has roots $r_1 = \frac{5}{2}$ and $r_2 = -\frac{3}{2}$. Let $g_1(x) = \frac{1}{2}(x^2 + x - \frac{15}{4})$ so that $g_1' = x + \frac{1}{2}$. It is instructive to draw the graph of $g_1(x)$.

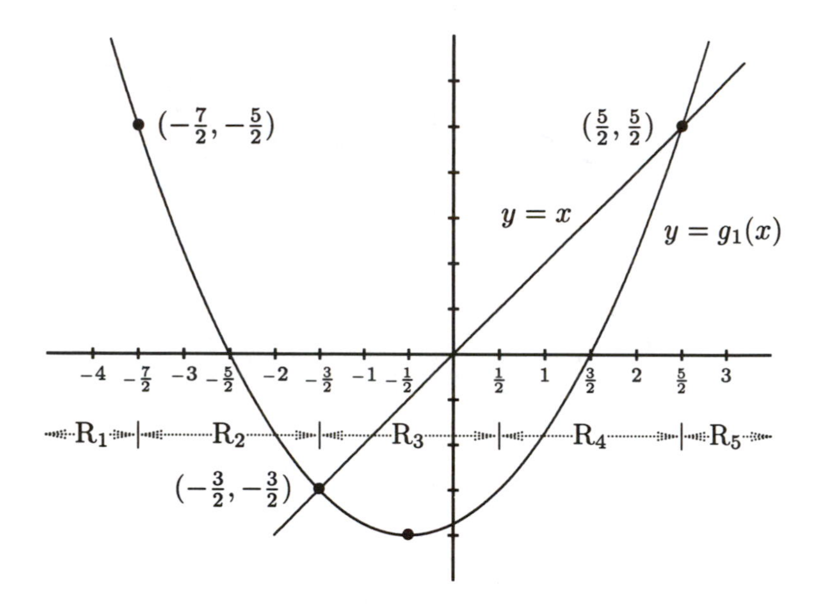

Figure 3.4

We first note that $r_1 = -\frac{3}{2}$ and $r_2 = \frac{5}{2}$ are fixed points of $g_1(x)$ since they are roots of $f(x)$. If $x_1 = \frac{1}{2}$, then $x_2 = -\frac{3}{2}$ which is now a fixed point. If $x_1 = -\frac{7}{2}$, then $x_2 = \frac{5}{2}$ which is also a root. For purposes of clarity we have divided the x-axis into five regions as pictured. If we start our iteration with $x_1 > \frac{5}{2}$ (in Region 5) then $g_1'(x) > 3$. By the Mean Value Theorem $(x_2 - \frac{5}{2}) = g_1(x_1) - g_1(\frac{5}{2}) = g_1'(\xi)(x_1 - \frac{5}{2}) > 3(x_1 - \frac{5}{2})$ so that $x_2 > x_1$. Continuing in this manner we see that the sequence generated has a limit of $+\infty$ as is seen in column 2 of Table 3.3.

If $x_1 < -\frac{7}{2}$ (in Region 1), then $x_2 = g_1(x_1)$ is in Region 5 and hence the generated sequence diverges to $+\infty$ as before. This is seen in column 3 of Table 3.3.

We next note that except in the special circumstances of starting with $x_1 = \frac{5}{2}$ or $x_1 = -\frac{7}{2}$, it is not possible for a generated sequence to converge to the root $r_2 = \frac{5}{2}$ since $g_1'(\frac{5}{2}) = 3 > 1$. We claim that if x_1 is in Region 2, 3 or 4 then the generated sequence converges to $r_1 = -\frac{3}{2}$ rather slowly. This claim is not immediately obvious since $g_1'(-\frac{3}{2}) = -1$ and we have neither the condition for convergence, $|g'(x)| \leq k < 1$, nor the condition of divergence, $|g'(x)| \geq k > 1$, in a neighborhood of $r_1 = -\frac{3}{2}$. In columns 4, 5 and 6 of Table 3.3 we have provided examples to illustrate this behavior. Note that $g_1'(x) = x + \frac{1}{2}$ so that $|g_1'(x)| < 1$ in Region 3. If we start in Region 2 or Region 4 we eventually end up in Region 3.

It shall be apparent to the reader that $g_1(x) = \frac{1}{2}\left(x^2 + x - \frac{15}{4}\right)$ is not a good iteration function for a multitude of reasons (such as divergence for many initial values and slowness of convergence even when it converges). We invite the reader to use $g_2(x) = \sqrt{x + \frac{15}{4}}$ for $x \geq -\frac{15}{4}$ to find the root $r_2 = \frac{5}{2}$ and $g_3(x) = -\sqrt{x + \frac{15}{4}}$ to find the root $r_1 = -\frac{3}{2}$. The reader may also use $g_4(x) = \frac{15/4}{x-1}$ to find the roots.

We close this section by noting that if $|g'(x)| > 1$ near a root r of $f(x)$ we may find this root by using the inverse function of g. If $h(g(x)) = x = g(h(x))$, then $h = g^{-1}$ is the inverse function of g and $x = g(x)$ implies that $h(x) = h(g(x)) = x$. Thus, if r is a fixed point of $g(x)$ and $|g'(x)| > 1$, then r is a fixed point of the inverse fuction $h(x)$ since $h(r) = h(g(r)) = r$ and $|h'(x)| < 1$.

As an example we consider the problem where $f(x) = x^2 - x - \frac{15}{4}$ with roots $r_1 = \frac{5}{2}$ and $r_2 = -\frac{3}{2}$ and iteration function $g_1(x) = \frac{1}{2}\left(x^2 + x - \frac{15}{4}\right)$ pictured above. In this case $g_1'(x) = x + \frac{1}{2}$ so that if $x_1 > \frac{5}{2}$ (Region 5), then $g'(x) > 3$ and the sequence $\{x_n\}$ such that $x_{n+1} = g_1(x_n)$ satisfies $x_n \to \infty$. If $h(x) = g^{-1}(x)$ for $x > \frac{5}{2}$ then the reader may verify, by solving $x = \frac{1}{2}\left(y^2 - y - \frac{15}{4}\right)$ for y, that $h(x) = -\frac{1}{2} + \sqrt{4 + 2x}$ and that $h'(x) = \frac{1}{\sqrt{4+2x}}$ so that $|h'(x)| < \frac{1}{3}$ for $x > \frac{5}{2}$. This last inequality is consistent with the chain rule

n	2	3	4	5	6
1	2.51	-3.6	-2.4	-1.5	2.49
2	2.53005	2.805	-0.195	-2	2.47005
3	2.5906015	3.4615125	-1.9534875	-0.875	2.4105985
4	2.77590881	5.84679064	-0.943687045	-1.9296875	2.23579181
5	3.36578926	18.14087571	-1.9015709	-1.97799683	1.74227841
6	5.4721633	171.741123	-1.01779951	-1.88575951	0.51390623
7	15.83336724	14831.5022	-1.84594183	-1.03983529	-1.48599708
8	131.389442	109994142	-1.06710146	-1.85428893	-1.51390488
9	8695.4124	6.04935565E+15	-1.83919797	-1.08295075	-1.48599845
10	37809444.23	1.829735189E+31	-1.1032744	-1.83008421	-1.51390353
11	7.14777054E+14	OVERFLOW	-1.81803	-1.115438	-1.48599982
12	2.554531184E+29	OVERFLOW	-1.13139846	-1.81061803	-1.51390218
13	OVERFLOW	OVERFLOW	-1.80066799	-1.14114019	-1.48600118
14	OVERFLOW	OVERFLOW	-1.15413139	-1.79446963	-1.51390084
15	OVERFLOW	OVERFLOW	-1.78605606	-1.16217419	-1.48500254
16	OVERFLOW	OVERFLOW	-1.17302991	-1.78076267	-1.5138995
17	OVERFLOW	OVERFLOW	-1.77351537	-1.17982349	-1.4850039
18	OVERFLOW	OVERFLOW	-1.1890793	-1.76892001	-1.51389815
19	OVERFLOW	OVERFLOW	-1.76258486	-1.194921	-1.48500527
20	OVERFLOW	OVERFLOW	-1.20293974	-1.7585424	-1.5138968
21	OVERFLOW	OVERFLOW	-1.75293786	-1.20803551	-1.48600664
22	OVERFLOW	OVERFLOW	-1.21507336	-1.74934286	-1.51389545
23	OVERFLOW	OVERFLOW	-1.74433504	-1.21957121	-1.48600801
24	OVERFLOW	OVERFLOW	-1.22581515	-1.74110864	-1.5138941
25	OVERFLOW	OVERFLOW	-1.73659618	-1.22982467	-1.48600938
26	OVERFLOW	OVERFLOW	-1.23541494	-1.73367798	-1.51389275
27	OVERFLOW	OVERFLOW	-1.72958243	-1.23901932	-1.48601075
28	OVERFLOW	OVERFLOW	-1.24406352	-1.72692522	-1.5138914
29	OVERFLOW	OVERFLOW	-1.72318474	-1.24732725	-1.48601211
30	OVERFLOW	OVERFLOW	-1.25190955	-1.72075099	-1.51389006
31	OVERFLOW	OVERFLOW	-1.71731601	-1.25488351	-1.48601347
32	OVERFLOW	OVERFLOW	-1.25917087	-1.71507544	-1.51388872
33	OVERFLOW	OVERFLOW	-1.71190571	-1.26179584	-1.48601483
34	OVERFLOW	OVERFLOW	-1.26564228	-1.70983355	-1.51388738
35	OVERFLOW	OVERFLOW	-1.70689595	-1.26815139	-1.48601619
36	OVERFLOW	OVERFLOW	-1.27170108	-1.70497172	-1.51388604
37	OVERFLOW	OVERFLOW	-1.70223872	-1.27402158	-1.48601755
38	OVERFLOW	OVERFLOW	-1.27731103	-1.7004453	-1.5138847
39	OVERFLOW	OVERFLOW	-1.69789378	-1.27946554	-1.48601891
40	OVERFLOW	OVERFLOW	-1.28252525	-1.69621674	-1.51388335

Table 3.3

which shows that if $h(g_1(x)) = x$ then $h'(g_1(x))g_1'(x) = 1$ so that $h'(g_1(x)) = \frac{1}{g_1'(x)}$. The values in the next paragraph show that if $x_1 = 5$ then the sequence $\{x_n\}$ converges to the root $r = \frac{5}{2}$ and that successive errors satisfy $x_{n+1} - 4 \approx \frac{1}{3}(x_n - r)$ if x_n is sufficiently close to r. We invite the reader to sketch the graph of $h(x)$ and $h'(x)$ for $x > \frac{5}{2}$ and recall that $h(x)$ is a reflection through the line $y = x$ of $g_1(x)$.

$x_1 = 5.0$	$x_6 = 2.508643919$	$x_{11} = 2.500035546$
$x_2 = 3.241657387$	$x_7 = 2.502879924$	$x_{12} = 2.500011849$
$x_3 = 2.737794739$	$x_8 = 2.500959821$	$x_{13} = 2.50000395$
$x_4 = 2.578244545$	$x_9 = 2.500319923$	$x_{14} = 2.500001316$
$x_5 = 2.525969116$	$x_{10} = 2.500106639$	$x_{15} = 2.500000439$

3.3 Newton's Method

The best known iteration method is called Newton's Method (or the Newton-Raphson Method). This method is motivated by the picture below and easily derived from the linear approximation given by Taylor series for $f(x_{n+1})$ about $x = x_n$. Thus, if we seek x_{n+1} so that $f(x_{n+1}) = 0$ and assume that our n^{th} iterate, x_n, satisfies $f'(x_n) \neq 0$ then the approximation

$$0 = f(x_{n+1}) \approx f(x_n) + (x_{n+1} - x_n)f'(x_n) \tag{3.6}$$

leads to the iteration step

$$x_{n+1} = x_n - \frac{f(x_n)}{f'(x_n)}.$$

If we define

$$g(x) = x - \frac{f(x)}{f'(x)} \tag{3.7}$$

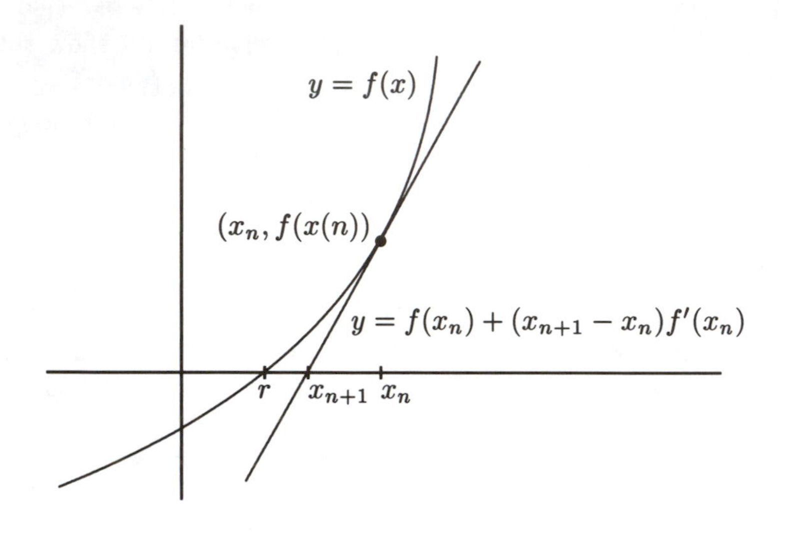

Figure 3.5

then Equation (3.6) is a special case of our study of iteration methods $x_{n+1} = g(x_n)$ begun in the previous section. To be precise we say that $g(x)$ given in (3.7) is the *Newton iteration* function for $f(x) = 0$, $x = g(x)$ is the Newton iteration equation for $f(x) = 0$, and that the corresponding iteration method is *Newton's method.*

As a simple example we apply Newton's method to find the root of our standard function

$$f(x) = x - \frac{3}{4}\sin x - \frac{2}{3}\,.$$

As will be noted in this table we started at the three different values $x_0 = 2$, which is near the root and $x_0 = 100$ and $x_0 = .01$, both of which are not near the root. In Exercise 3.10 we ask the reader to show that in this special example Newton's method converges for any starting value to a unique root.

Table 3.4 was obtained by applying Newton's method to our standard example with the three starting values 2, 100 and 0.1. We will see in the next paragraph the fact that all three sequences converge to the root r is a pleasant and unexpected circumstance not easily established, when $x_1 = 100$ for example, by (3.8).

n	x_n	x_n	x_n
1	2	100	0.01
2	1.50357804	-182.264701	2.66626772
3	1.41027262	546.314413	1.67255375
4	1.40658205	-1344.287162	1.43117659
5	1.4065763	3345.763804	1.40682692
6	1.4065763	1433.864648	1.40657632
7	1.4065763	-360.945423	1.4065763
8	1.4065763	-149.324133	1.4065763
9	1.4065763	13.413546	1.4065763
10	1.4065763	-10.79041622	1.4065763
11	1.4065763	-0.21491213	1.4065763
12	1.4065763	2.48526656	1.4065763
13	1.4065763	1.63157235	1.4065763
14	1.4065763	1.42470552	1.4065763
15	1.4065763	1.40671306	1.4065763
16	1.4065763	1.40657631	1.4065763
17	1.4065763	1.40657629	1.4065763
18	1.4065763	1.40657629	1.4065763

Table 3.4

Problem 3.3: Use Table 3.4 to sketch $f(x) = x - \frac{3}{4}\sin x - \frac{2}{3}$ at $x = 100$ and $x = -182$.

Newton's method has the highly desirable property that for simple roots r of $f(x) = 0$, we can start close to r and generate the sequence $\{x_n\}$ by using (1) so that this sequence converges to r. To see this we note that if $g(x)$ is as in (3.7) then

$$g'(x) = 1 - \frac{f'^2(x) - f(x)f''(x)}{f'(x)^2} = \frac{f(x)f''(x)}{f'(x)^2} \tag{3.8}$$

so that $g'(r) = 0$ and by continuity there is an interval $I = (r-c, r+c)$ so that $|g'(x)| \leq k < 1$ holds for x in I. In particular, if x_1 is in I then x_n is in I for each I and the reasoning in the previous section shows that $x_n \to r$.

A second desirable property of Newton's method is that the convergence is quadratic. As we will soon see if our starting value x_1 is close enough to a simple root r of $f(x)$, then the sequence $\{x_n\}$ generated by $x_{n+1} = g(x_n)$ converges *quadratically* to r, which means

$$\lim_{n\to\infty} \frac{|x_{n+1} - r|}{|x_n - r|^2} = \text{constant} > 0\,. \tag{3.9}$$

Intuitively, this means that (in the limit) each iteration step doubles the number of significant digits.

Problem 3.4: Show carefully that the convergence for $f(x) = x - \frac{3}{4}\sin x - \frac{2}{3}$ is quadratic.

Problem 3.5: Generate the first column of Table 3.4 obtaining 12 place accuracy and decide if each iteration step eventually doubles the number of significant digits.

A common misuse of Newton's method is that the user often doesn't start close enough to the root r. The reader can see from (3.8) that this can easily happen if $f'(x)$ is small or if $f''(x)$ is large. This misuse occurs in real life and in "cookbook" courses in numerical analysis and results in loss of large amounts of computer time as well as real time.

A very illustrative and practical exercise is to use Newton's method to find a square root of $c > 0$ (say $c = 5$). The reader should pay careful attention to this example since it also illustrates why Newton's method converges quadratically near the root.

We define $f(x) = x^2 - 5$ so that

$$g(x) = x - \frac{f(x)}{f'(x)} = x - \frac{x^2-5}{2x} = \frac{x^2+5}{2x} = \frac{1}{2}\left(x + \frac{5}{x}\right). \tag{3.10}$$

Equation (3.9) is the basis of the high school rule

$$x_{n+1} = \frac{1}{2}\left(x_n + \frac{c}{x_n}\right)$$

for finding $\sqrt{c}$ for $c > 0$ which was used in prehistoric times (before hand calculators). In fact, this method is so efficient that it is used in hand calculators. To see why this rule works so well we observe, as above, that

$$\begin{aligned} g'(x) &= 1 - \frac{[f'(x)^2 - f(x)f''(x)]}{f'(x)^2} = \frac{f(x)f''(x)}{f'(x)^2} \\ &= \frac{(x^2-5)(2)}{(2x)^2} = \frac{1}{2}\left(1 - \frac{5}{x^2}\right) \end{aligned} \tag{3.11}$$

so that $g'(r) = g'(\sqrt{5}) = 0$. By continuity, there is a neighborhood at $r = \sqrt{5}$ so that the condition $|g'(x)| \leq k < 1$ holds to this neighborhood.

To justify our remarks about quadratic convergence in the general case we use a Taylor series expansion about r and note that $r = g(r)$, $g'(r) = 0$ and $x_{n+1} = g(x_n)$. Thus,

$$\begin{aligned} x_{n+1} - r &= g(x_n) - g(r) \\ &= (x_n - r)g'(r) + \frac{(x_n - r)^2}{2} g''(\xi_n) \end{aligned} \tag{3.12a}$$

or

$$\frac{|x_{n+1} - r|}{|x_n - r|^2} = \frac{1}{2}|g''(\xi_n)|, \tag{3.12b}$$

where ξ_n is between x_n and r. Since $x_n \to r$, $\xi_n \to r$ so that we may choose the constant in (3.9) as

$$\lim_{n\to\infty} \frac{1}{2}|g''(\xi_n)| = \frac{1}{2}|g''(r)|\,.$$

In our particular example $g''(x) = \frac{5}{x^3}$ so that $\frac{1}{2}|g''(r)| = \frac{1}{2\sqrt{5}}$.

A complete analysis when $r = \sqrt{5}$ can be made using elementary calculus. For $x > 0$, our picture is as below. It can be easily shown that $|g'(x)| < 1$ for $x > \sqrt{5/3}$. By (3.12b), there exists an $M > 0$ such that $|x_{n+1} - \sqrt{5}| \le M|x_n - \sqrt{5}|^2$ if $x > \sqrt{5/3}$. We leave it to the reader to show that $M = \frac{3\sqrt{3}}{2\sqrt{5}}$ if $x > \sqrt{5/3}$. Thus, any initial guess $x_1 > \sqrt{5/3}$ leads to a sequence $\{x_n\}$ which converges quadratically to $\sqrt{5}$. If $0 < x_1 \le \sqrt{5/3}$ or, more generally. if $0 < x_1 < \sqrt{5}$ then $g'(x) < 0$ so that $x_2 = g(x_n) > \sqrt{5}$ and our sequence still converges quadratically.

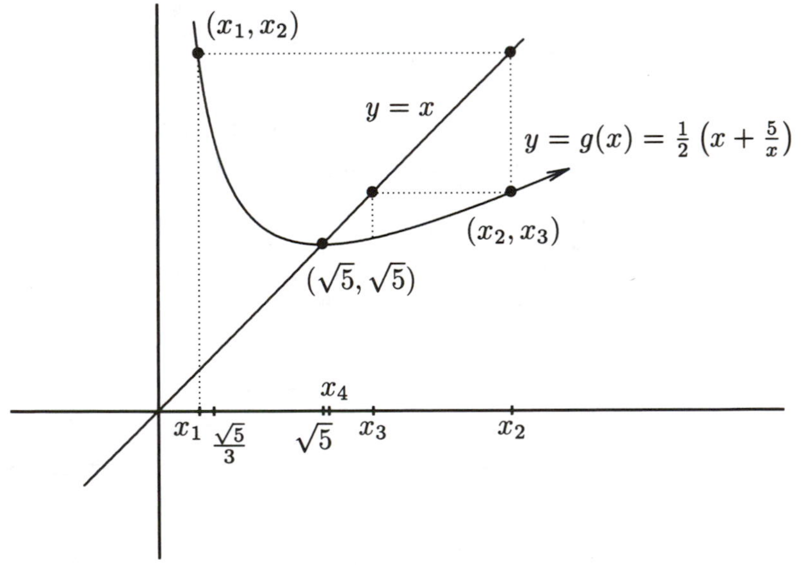

Figure 3.6

In Exercise 3.15 we will see that the square root algorithm will converge for any positive starting value $x_1 > 0$. The following table was obtained by applying Newton's method to the function $f(x) = x^2 - 5 = 0$ with the three starting values 2, 100 and 0.01.

n	x_n	x_n	x_n
1	2	100	0.01
2	2.25	50.025	250.005
3	2.23611112	25.06247502	125.0125
4	2.23606798	12.63098824	62.526249
5	2.23606798	6.51342008	31.30310771
6	2.23606798	3.64053298	15.73141814
7	2.23606798	2.50697913	8.02462675
8	2.23606798	2.25070569	4.32385435
9	2.23606798	2.23606798	2.74011502
10	2.23606798	2.23606798	2.28242798
11	2.23606798	2.23606798	2.23653881
12	2.23606798	2.23606798	2.23606803
13	2.23606798	2.23606798	2.23606798
14	2.23606798	2.23606798	2.23606798

Table 3.5

To illustrate several important ideas, we consider the function $f(x) = \log x - 2$ which is defined for $x > 0$ and has a unique root $r = e^2$. The Newton iteration function is

$$g_1(x) = x - \frac{\log x - 2}{1/x} = 3x - x \log x$$

with

$$g_1'(x) = \frac{f(x)f''(x)}{f'(x)^2} = 2 - \log x \,.$$

If $x_1 \geq e^3$, the reader should verify that $x_2 < 0$ so that our iteration terminates. If $e \leq x_1 < e^3$ the reader should repeat the methods given above to show that the iteration converges quadratically to $r = e^2$.

On the other hand, if we use $g_2(x) = x - \log x + 2$, our iteration converges for any starting value $x_1 > 0$. The reader should verify this fact. However, this convergence is very slow, since $g'(x) = 1 - \frac{1}{x} > 0.86$ for x close to $r = e^2$ and $g'(x) \approx 1$ for large values of x. For example, if $x_1 = 100$ then $x_{100} = 7.396396$ which is correct to only 3 significant figures.

Problem 3.6: Sketch $f(x) = x - \log x + 2$ and justify all statements in the last paragraph.

3.4 Secant Method

The purpose of this brief section is to describe a method of iteration known as the secant method. This method is often used in place of Newton's method when it is impossible or impractical to formally differentiate $f(x)$ at the iteration points x_n. The idea is to allow the values of the function $f(x)$ at x_n and x_{n-1} to approximate the values $f'(x_n)$ needed for Newton's method.

The formula is easy to derive. Let $f'(x_n)$ in Newton's method be approximated by

$$f'(x_n) \approx \frac{f(x_n) - f(x_{n-1})}{x_n - x_{n-1}}.$$

Then from Newton's method we have

$$x_{n+1} = x_n - \frac{f(x_n)}{f'(x_n)} \approx x_n - f(x_n)\left[\frac{x_n - x_{n-1}}{f(x_n) - f(x_{n-1})}\right].$$

Thus, if x_1 and x_2 are given the sequence $\{x_n\}$ is determined by

$$x_{n+1} = x_n - \left[\frac{x_n - x_{n-1}}{f(x_n) - f(x_{n-1})}\right] f(x_n)\,.$$

This method is called the *secant method.*

This method gets its name from the fact that we obtain x_{n+1} by drawing the secant line through $(x_n, f(x_n))$ and $(x_{n-1}, f(x_{n-1}))$ as in Figure 3.7, below. Unlike the previous two iteration methods, the secant method depends on the two previous terms of sequence. Thus, given the points x_1 and x_2 we will be able to generate the sequence $x_3, x_4, \ldots$

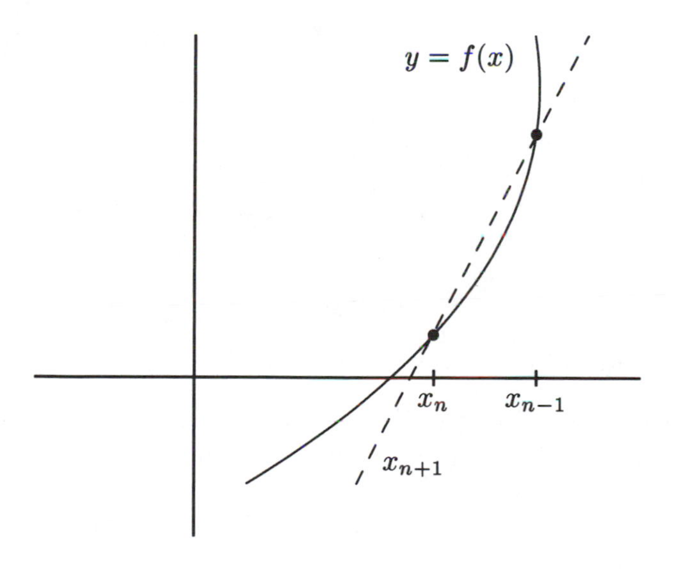

Figure 3.7

A practical disadvantage of this method is that if f is continuous, which we are assuming, and the x_n are converging to a root of $f(x) = 0$, then $f(x_n) - f(x_{n-1})$ could be approaching zero and hence cause computer overflow. Also if $f(x_n)$ and $f(x_{n-1})$ are of the same sign, then catastrophic cancellation is also a possibility. Thus we may want to stop the iteration procedure when

$$|f(x_n) - f(x_{n-1})| \leq \epsilon |f(x_n)|$$

for some preassigned $\epsilon > 0$. Finally, we note that the secant method has the advantage of requiring only one new function evaluation per iteration, unlike Newton's method which requires two.

It can be shown that the secant method has a convergence rate which is between linear and quadratic. In fact, if $e_n = |x_n - r|$, then it can be shown that

$$e_{n+1} \leq C\, e_n^{(1+\sqrt{5})/2}$$

where C is a constant and $\frac{1+\sqrt{5}}{2} \approx 1.62$.

We close this section by applying the secant method to our standard example, that is, we wish to find the root of the equation

$$f(x) = x - \frac{3}{4}\sin x - \frac{2}{3} = 0\,.$$

The reader may also apply this method to any of the examples for Newton's method. It might be noted that for the far away starting values, namely, $x_1 = 100$ and $x_2 = 101$, that we get faster convergence than when we started Newton's method with $x_1 = 100$. However, when we started near the root Newton's method was faster.

The following table was obtained by applying the secant method to the function

$$f(x) = x - \frac{3}{4}\sin x - \frac{2}{3}$$

with the pairs of starting values $x_1 = 1$, $x_2 = 2$ and $x_1 = 100$, $x_2 = 101$.

n	x_n	x_n
1	1	100
2	2	101
3	1.31372935	-254.589707
4	1.38736667	0.925682
5	1.40735979	1.2710936
6	1.4065699	1.44006985
7	1.40657628	1.40459153
8	1.40657628	1.40654858
9	1.40657628	1.40657631
10	1.40657628	1.40657629

Table 3.6

Problem 3.7: Generate the first column of Table 3.6 obtaining 12 place accuracy and try to decide if the convergence is (slightly) less than in Problem 3.5.

3.5 Theoretical Ideas

In this section we will present theoretical ideas which explain why the intuitive ideas of the last two sections work. The importance of these ideas can not be overly emphasized.

Theorem 3.1 is called an existence theorem while Theorem 3.2 is a uniqueness theorem. Make sure you understand both the hypothesis and conclusion of each theorem. For example, in Theorem 3.1 the "if" part is the hypothesis and the "then" part is the conclusion. In Theorem 3.2, parts (i), (ii) and (iv) are the hypotheses while parts (iii) and (v) are the conclusions.

Theorem 3.1: *If $g(x)$ is continuous and maps the interval $[a, b]$ into $[a, b]$ then $g(x)$ has (at least) one fixed point r on $[a, b]$.*

Proof: If $a = g(a)$ or $b = g(b)$ we are done. If neither of these equalities hold let $h(x) = x - g(x)$. Now $h(a) = a - g(a) < 0$ and $h(b) = b - g(b) > 0$. The the Intermediate Value Theorem, there exists (at least one) x in (a, b) such that $h(x) = 0$ and hence that $x = g(x)$.

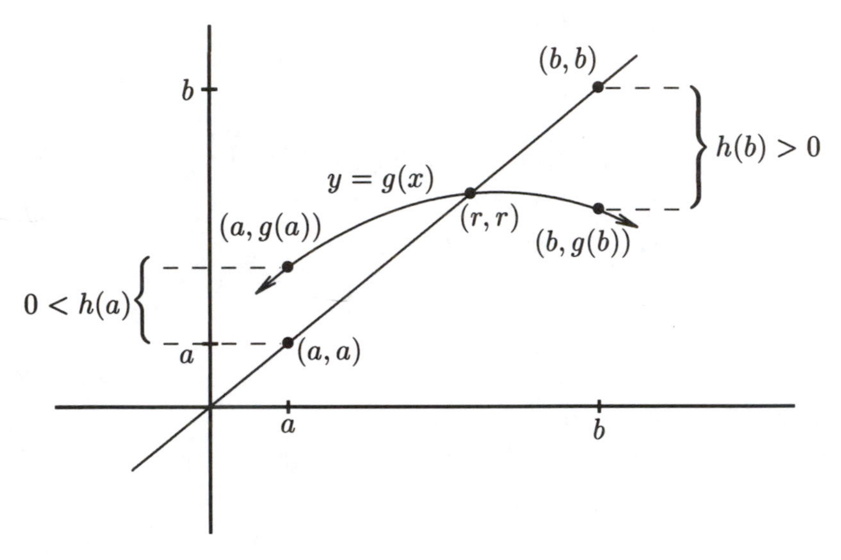

Figure 3.8

Theorem 3.2: *Assume that* (i) $g(x)$ *is continuous on* $[a, b]$ *and maps the interval* $[a, b]$ *into* $[a, b]$, *and* (ii) $g'(x)$ *exists on* $[a, b]$ *and* $|g'(x)| \leq k < 1$, *for some constant* k, *on* $[a, b]$. *Then,* (iii) *there exists a unique fixed point* r *of* $g(x)$ *in* $[a, b]$ *and* (iv) *if* x_1 *is any point in* $[a, b]$, *then* (v) *the sequence* $\{x_n\}$ *defined by* $x_{n+1} = g(x_n)$ *converges to* r.

Proof: From (i) and Theorem 3.1, there is at least one fixed point r of $g(x)$ in $[a, b]$. Suppose there exist at least two distinct fixed points r_1 and r_2 of $g(x)$ on $[a, b]$. Then $r_1 = g(r_1)$ and $r_2 = g(r_2)$

and there exists ξ between r_1 and r_2 such that

$$1 = \frac{r_2 - r_1}{r_2 - r_1} = \frac{g(r_2) - g(r_1)}{r_2 - r_1} = g'(\xi) \leq k < 1\,.$$

The contradiction $1 < 1$ shows that there is at most one fixed point r in $[a, b]$ and (iii) is established.

To obtain (v) we use the Mean Value Theorem to obtain $x_{n+1} - r = g(x_n) - g(r) = g'(\xi)(x_n - r)$ for some ξ between x_n and r. Thus $|x_{n+1} - r| \leq k|x_n - r|$. Repeated application of this last result gives $|x_{n+1} - r| \leq k|x_n - r| \leq k^2|x_{n-1} - r| \leq \cdots \leq k^n|x_1 - r| \leq k^n|b - a|$ so that $x_{n+1} - r \to 0$ or $x_{n+1} \to r$, the unique fixed point.

Corollary 3.3 is a practical result since we know the values of k, x_1 and x_2 and hence we can put a bound on $|x_n - r|$. Note that the smaller k is, the better the value of $|x_n - r|$.

Corollary 3.3: *Under the hypothesis in Theorem 3.2,*

$$k|x_n - r| \leq \frac{k^{n-1}}{1-k}|x_2 - x_1|\,.$$

Proof: As above we have $|x_{n+1} - x_n| \leq k|x_n - x_{n-1}| \leq \cdots \leq k^{n-1}|x_2 - x_1|$. Thus, for $m > n \geq 1$, we have by the triangular inequality

$$\begin{aligned} |x_m - x_n| &= |x_m - x_{m-1} + x_{m-1} - x_{m-2} + \ldots + x_{n+1} - x_n| \\ &\leq |x_m - x_{m-1}| + |x_{m-1} - x_{m-2}| + \ldots + |x_{n+1} - x_n| \\ &\leq k^{m-2}|x_2 - x_1| + k^{m-3}|x_2 - x_1| + \ldots + k^{n-1}|x_2 - x_1| \\ &= k^{n-1}|x_2 - x_1|\big(k^{m-n-1} + k^{m-n-2} + \ldots + k + 1\big) \\ &\leq \frac{k^{n-1}}{1-k}|x_2 - x_1|\,. \end{aligned}$$

This last step follows since if $0 < k < 1$ the infinite geometric series $1 + k + k^2 + \ldots$ is bounded by $\frac{1}{1-k}$. Finally, fixing n and noting that

$x_m \to r$, we have the desired result, since $|x_m - x_n| = |x_n - x_m| \to |x_n - r|$.

3.6 Real Roots of Polynomials

In this section we shall consider the problem of finding real roots of a polynomial

$$P(x) = a_n x^n + a_{n-1} x^{n-1} + \cdots + a_1 x + a_0 \,, \tag{3.13}$$

where the coefficients a_k, $k = 0, 1, \ldots, n$, are real numbers and $n \geq 1$. We present two fundamental results on the problem of finding roots of a polynomial.

Theorem 3.4: (Fundamental Theorem of Algebra). *The polynomial $P(x)$ as defined in (3.13), has at least one root, which might be complex. That is, there exists a (possibly complex) number r such that $P(r) = 0$.*

Theorem 3.5: (Factor Theorem). *A number r is a root of the polynomial $P(x)$, as defined in (3.13), if and only if there exists a polynomial of degree $n-1$, say $Q(x)$, such that*

$$P(x) = (x - r)Q(x) \,.$$

A corollary of these results is the following result.

Corollary 3.6: *For the polynomial $P(x)$, defined in (3.13), there exist numbers $x_1, \ldots, x_m$ (possibly complex) and integers $n_1, \ldots, n_m$ such that $n_1 + \cdots + n_m = n$ and*

$$P(x) = a_n (x - x_1)^{n_1} (x - x_2)^{n_2} \cdots (x - x_m)^{n_m} \,.$$

It should be stated these three results also hold for polynomials with complex coefficients, but since we will not be considering the case of polynomials with complex coefficients we stated the results only for polynomials with real coefficients.

We see from these results that if we can find a root, then we keep dividing out the factors according to its multiplicity to get a lower degree polynomial on which to work. This process is call *deflation.* To do this division easily we shall now discuss synthetic division or Horner's method, as it is often called.

Theorem 3.7: (Horner). *Let $P(x)$ be defined as in* (3.13), *let x_0 be a given real number and consider the sequence $b_0, \ldots, b_n$ defined by*

$$\begin{aligned} b_n &= a_n \\ b_k &= a_k + b_{k+1}x_0, \quad k = n-1, n-2, \ldots, 1, 0 \,. \end{aligned} \tag{3.14}$$

Then $b_0 = P(x_0)$ and if

$$Q(x) = b_n x^{n-1} + b_{n-1}x^{n-2} + \cdots + b_2 x + b_1 \,,$$

then

$$P(x) = (x - x_0)Q(x) + b_0 \,.$$

This theorem follows easily by simply multiplying out $(x - x_0)Q(x) + b_0$ and using the recurrence (3.14).

Horner's method is often given in the form of nested multiplication, that is,

$$P(x) = a_0 + x(a_1 + x(a_2 + \cdots + x(a_{n-1} + a_n x) \cdots)) \,.$$

This representation of the polynomial $P(x)$ is better than that in (3.13) if we wish to evaluate P at a specific value x_0 in that the *round-off* errors which often occur in computing the powers of x_0 are avoided.

To illustrate this theorem let

$$P(x) = 3x^4 - 3x^3 + 2x^2 - 4 \tag{3.15}$$

and let $x_0 = 2$. Then

$$b_4 = 3, \quad b_3 = -3 + 3(2) = 3, \quad b_2 = 2 + 3(2) = 8,$$
$$b_1 = 0 + 8(2) = 16 \quad b_0 = -4 + 16(2) = 28\,.$$

Hence $P(2) = 28$ and

$$P(x) = (x-2)(3x^3 + 3x^2 + 8x + 16) + 28\,.$$

The name synthetic division is often applied to this procedure when it is written in "tabular" form. This example would be presented as:

2⌋	3	-3	2	0	-4
		6	6	16	32
	3	3	8	16	28

This gives us the same numbers as above: the first four numbers are the coefficients of $Q(x)$, in descending order and the last number is the remainder, which is $P(2)$.

One thing that should be noted in this procedure is that if

$$P(x) = (x - x_0)Q(x) + b_0\,,$$

then

$$P'(x) = Q(x) + (x - x_0)Q'(x)$$

so that

$$P'(x_0) = Q(x_0)\,.$$

Thus, if we use Newton's method to find the roots of the polynomial $P(x)$, then we can use the same procedure to calculate both P and P'.

We illustrate this by finding the first few approximations to the root of $P(x)$, defined in (3.15), which lies in the interval $[1, 2]$. Suppose we use $x_1 = 2$ as a starting value in the Newton iteration. Then

$$x_2 = x_1 - \frac{P(x_1)}{P'(x_1)} .$$

Above we found that $P(2) = 28$ and $Q(x) = 3x^3 + 3x^2 + 8x + 16$. In a similar manner we can calculate $Q(2)$:

2⌋	3	3	8	16
		6	18	52
	3	9	26	68

so that $Q(2) = 68$. Since $P'(2) = Q(2) = 68$ we have

$$x_2 = 2 - \frac{28}{68} = \frac{27}{17} \approx 1.588 .$$

Now we repeat the process with x_2 in place of x_1. We have

1.588⌋	3.	-3.	2.	0.	-4.	
		4.764	2.801	7.624	12.107	
1.588⌋	3.	1.764	4.801	7.624	8.107	$= P(x_2)$,
		4.764	10.366	24.085		
	3.	6.528	15.167	31.709	$= Q(x_2) = P'(x_2)$	

Thus $P(1.588) = 8.107$ and $P'(1.588) = 58.461$ and

$$x_3 = 1.588 - \frac{8.107}{31.461} \approx 1.330 .$$

To find further approximations we simply repeat the process.

We note that the polynomial $Q(x)$ changes with each iteration. However, it is still completely determined by $P(x)$ and the iterate x_N. Once we have an iterate x_N that is an approximate root of $P(x)$, that is, $P(x_N)$ is sufficiently small, then we have

$$P(x) = (x - x_N)Q(x) + P(x_N) \approx (x - x_N)Q(x),$$

that is, $x - x_N$ is an approximate factor of $P(x)$. If we let $y_1 = x_N$, then we have

$$P(x) \approx (x - y_1)Q_1(x).$$

Now we use $Q_1(x)$ as the primary polynomial and use Newton's method, as outlined above, to find the roots of $Q_1(x)$. We repeat this process until we get to a polynomial quotient $Q_J(x)$, which is either a quadratic polynomial, to which we apply the quadratic formula and hence obtain the last two approximate roots of $P(x)$, or is a polynomial we can show has no real roots.

A problem with this method is that it is based on many approximations. As we saw in Chapter 2, perturbing even one coefficient can be enough to completely change the structure of the roots. Thus the roots of $Q_1(x)$, for example, may have nothing to do with the roots of $P(x)$.

One way to get around this problem is to iterate the approximate roots of $Q_1(x)$ in $P(x)$. That is, if x_N is the approximate root of $Q_1(x)$ we have found we take as our next approximate root of $P(x)$

$$y_2 = x_N - \frac{P(x_N)}{P'(x_N)}$$

and then find the next quotient $Q_2(x)$ by applying synthetic division to $P(x)$.

We illustrate this by continuing the root finding of the polynomial $P(x)$ defined in (3.15). We have

$$P(x) \approx (x - 1.330)(3x^3 + .990x^2 + 3.317x + 4.412)$$

so that

$$Q_1(x) = 3x^3 + .990x^2 + 3.317x + 4.412\,.$$

Now $P(x)$ also has a root in the interval $[-1, 0]$ so we shall take as our first approximation to the root of $Q_1(x)$, $x_1 = -1$. We have

-1	3.	0.990	3.317	4.412
		-3.	2.01	-5.327
-1	3.	-2.01	5.327	-0.915 $= Q_1(x_1)$
		-3.	5.01	
	3.	-5.01	10.337 $= Q_1'(x_1)$	

Thus

$$x_2 = -1 - \frac{(-0.915)}{10.337} \approx -0.911\,.$$

Suppose we decided this iteration was a good approximation of the root to $Q_1(x)$. Then we would compute

$$x_3 = x_2 - \frac{P(x_2)}{P'(x_2)} \approx -0.911 - \frac{1.995}{-20.187} \approx -0.812$$

and use this value as our y_2. Thus we have $y_1 = 1.330$ and $y_2 = -0.812$. To find $Q_2(x)$ we use synthetic division on $P(x)$.

-0.812	3.	-3.	2.	0.	-4.
		-2436	-1.978	-0.018	0.01965
1.330	3.	-52.436	0.022	-0.018	-3.985
		3.99	-1.923	-2.528	
	3.	-1.446	-1.901	-2.546	

Thus

$$Q_2(x) = 3x^2 - 1.446x - 1.901\,.$$

It might be noted that our approximate roots y_1 and y_2 are not really that good (the two remainders -3.985 and -2.546 are quite large), but we only wished to illustrate the procedure. It might also be noted that $Q_2(x)$ is a quadratic polynomial to which we could apply the quadratic formula to find its roots. At the end of this section we shall completely do one example to illustrate the material discussed in this section.

Since we are using Newton's method to find the roots we know that, in general, we must start relatively close to the root to guarantee convergence of the sequence formed from the iteration process. Thus the question arises: how do we find our initial guess x_1? We can do this, proceeding somewhat as we did in the bisection method, namely, find two values x_1 and x_2 so that $P(x_1)P(x_2) \leq 0$, where we can evaluate $P(x_1)$ and $P(x_2)$ by Horner's method. The Intermediate Value Theorem tells us that in this case $P(x)$ must have at least one root between x_1 and x_2. In fact one might wish to begin the process of finding the roots of $P(x)$ by using a few iterations of the bisection method to narrow the interval down before beginning the use of Newton's method.

Before we close this section with our final example, we wish to state some facts that are often useful in deciding where to look for the roots. The first result is a result ascribed to R. Descartes.

Theorem 3.8: (Rule of Signs). *Write down, in order, the signs of the nonzero coefficients of the polynomial. Count the number of sign changes in the list, say S. Let k be the number of positive roots. Then $k \leq S$ and $S - k$ is a nonnegative even integer.*

An immediate corollary allows us to get an estimate of the number of negative roots of a polynomial: If we count the number of sign changes in the coefficients of $P(-x)$, then the number of negative roots is within an even integer of the number of sign changes. As an example let

$$P(x) = x^4 - 2x^3 + x^2 + 8x - 2\,.$$

Then the number of sign changes is 3. Thus there is either 1 or 3 positive roots. Also

$$P(-x) = x^4 + 2x^3 + x^2 - 8x - 2\,.$$

Here the number of sign changes is 1. Thus there is exactly one negative root since the number of sign changes in the coefficients of $P(-x)$ insures that the number of negative roots must be within a nonnegative even integer of 1, and so this even integer must be zero in this case.

We next consider two results that are useful when doing synthetic division.

Theorem 3.9: *Let $P(x)$ be defined in (3.13) with $a_n > 0$ and let the numbers $\{b_k\}$ be as in Theorem 3.7.*

1. *If x_0 is a positive number such that $P(x_0) > 0$ and $b_k > 0$ for $k = 1, 2, \ldots, n$, then all of the roots of P are less than or equal to x_0.*
2. *If x_0 is a negative number such that $(-1)^n P(x_0) > 0$ and the b_k alternate in sign (that is, $b_n > 0$, $b_{n-1} < 0$, $b_{n-2} > 0$, etc.), then all the roots of P are greater than or equal to x_0.*

Both of these results follow easily from Theorem 3.7. For, by Theorem 3.4, we have

$$P(x) = (x - x_0)(b_n x^n + \cdots + b_1) + P(x_0)\,.$$

For the first part of the theorem, note that if $x > x_0 > 0$, $P(x_0) > 0$ and $b_k > 0$, $k = 1, \ldots, n$, then $P(x) > 0$ since it is the product of two positive numbers added to another positive number. Thus x cannot be a root. The second part follows in a similar manner.

One last fact concerns the validity of the approximation process. The process of synthetic division used to derive the new polynomial $Q(x)$, whose roots we now wish to find, has coefficients which are only approximations to the correct coefficients since our roots are only approximations. It can be shown that if we determine the roots in order of increasing absolute value and with the maximum allowable accuracy we can avoid the problem of finding successive roots (to the successive $Q(x)$ polynomials) that are not roots to the original polynomial.

We close this section by determining the roots, to an accuracy of four decimal places, of the polynomial

$$P(x) = x^4 - 3.6x^3 + 0.17x^2 + 6.57x - 0.6552\,.$$

By Descarte's Rule of Signs we see that we have 1 or 3 positive roots and 1 negative root. Thus there are at least two roots that we should be able to locate and can then use the quadratic formula on the resulting quadratic polynomial formed by dividing out the factors corresponding to these roots.

Before beginning the Newton method we attempt to find some intervals containing these roots. We use intervals of length 0.5 and obtain the following table.

$P(-2.5) =$	79.2948	$P(0.5) =$	2.2848
$P(-2.0) =$	31.6848	$P(1.0) =$	3.4848
$P(-1.5) =$	7.0848	$P(1.5) =$	2.4948
$P(-1.0) =$	-2.4552	$P(2.0) =$	0.3648
$P(-0.5) =$	-3.3852	$P(2.5) =$	-0.3552
$P(0.0) =$	-0.6552	$P(3.0) =$	4.3848

Table 3.7

Thus we see, by the Intermediate Value Theorem, that we have at least one root in each of the intervals $(-1.5, -1.0)$, $(0, .5)$, $(2.0, 2.5)$ and $(2.5, 3.0)$. By Corollary 3.6 the polynomial, being of degree four, only has four roots. Thus, each of the intervals contains exactly one root of the polynomial. By the remark made above about the order in which to find the roots we should try to find the root in the interval $(0, .5)$ first, then the one in $(-1.5, -1.0)$ and then we'll have a quadratic polynomial to which we can apply the quadratic formula.

We first approximate the root in the interval $(0, 0.5)$. We shall take $x_1 = 0$. Then we have

0	1.	-3.6	0.17	6.57	-0.6552
		0.	0.	0.	0.
0	1.	-3.6	0.17	6.57	-0.6552 $= P(0)$
		0.	0.	0.	
	1.	-3.6	0.17	6.57 $= P'(0)$.	

Thus

$$x_2 = 0 - \frac{-.6552}{6.57} = .0997.$$

If we continue using synthetic division, as above, to compute $P(x_n)$ and $P'(x_n)$ and then put these values in the Newton method iteration scheme we obtain the following table.

N	x_N	$P(x_N)$	$P'(x_N)$
1	0.0000	-0.6552	6.5700
2	0.0997	-0.0019	6.5006
3	0.1000	0.0000	6.5

Table 3.8

Thus, we take as our approximation to the smallest root in absolute value $y_1 = .1$. Then, using synthetic division we see that

$$P(x) = (x - .1)(x^3 - 3.5x^2 - .18x + 6.552).$$

We next approximate the root in the interval $(-1.5, -1.0)$. We take as our first approximation $x_1 = -1.5$. The table for this root is as follows.

N	x_N	$P(x_N)$	$P'(x_N)$
1	-1.5	-4.428	17.07
2	-1.2406	-0.5209	17.1215
3	-1.2009	-0.0113	17.5528
r	-1.2000	0.0000	17.5400

Table 3.9

Thus, we take as our approximation to the second smallest root in absolute value $y_2 = -1.2$. Then, using synthetic division we see that

$$P(x) = (x - .1)(x + 1.2)(x^2 - 4.7x + 5.46)\,.$$

To find the last two roots we can use the quadratic formula and we find that the last two roots are $y_3 = 2.1$ and $y_4 = 2.6$. Thus the polynomial $P(x)$ has the four roots

$$y_1 = .1, \quad y_2 = -1.2, \quad y_3 = 2.1 \quad \text{and} \quad y_4 = 2.6\,,$$

which, indeed, are the exact roots.

Exercise Set 3

1. Suppose $f(x)$ has only simple roots in (a, b). If $f(a)f(b) < 0$, show that there are an odd number of roots of $f(x) = 0$ in (a, b). If $f(a)f(b) > 0$, show that there are an even number (possibly zero) of roots of $f(x) = 0$ in (a, b).

2. Show that the bisection method converges linearly, that is, $\lim_{n\to\infty} \frac{|c_{n+1}-r|}{|c_n - r|}$ is a constant.

3. Show that in the bisection method, if one wants an accuracy of ϵ in their answer, that is, the iteration is stopped when

$$a_{n+1} - a_n \le \epsilon\,,$$

then the number of steps necessary to achieve this is no more than

$$1 + \frac{\log \frac{b-a}{\epsilon}}{\log 2}\,.$$

4. Suppose that g is differentiable on $[a, b]$ and satisfies $|g'(x)| \le k < 1$, for all x in $[a, b]$. Then the equation $x = g(x)$ has at most one root in the interval $[a, b]$. (Hint: Use the Mean Value Theorem.)

5. Suppose g is differentiable on $[a, b]$ and that the equation $x = g(x)$ has a root in $[a, b]$. Suppose g satisfies the following conditions on $[a, b]$:

   a) $a < g(a)$ and $b > g(b)$ and

   b) g is increasing on $[a, b]$.

   If x_1 is in $[a, b]$ and we define the sequence $x_{n+1} = g(x_n)$, for $n \ge 1$, then the sequence $\{x_n\}$ converges to a unique number ξ in

(a, b) such that $\xi = g(\xi)$. Moreover, if $x_1 < \xi$, then the sequence is monotone increasing. (Hint: Show that for $a \le x < \xi$ we have $x - g(x) < 0$ and for $\xi < x \le b$, $x - g(x) > 0$.)

6. Suppose g is differentiable on $[a, b]$ and has a root in $[a, b]$. Suppose that for all x in $[a, b]$ there is a constant k such that $|g'(x)| \le k < 1$. Then the sequence $\{x_n\}$ defined in Exercise 5 converges to the unique root of $x = g(x)$. (Hint: Let $\omega(x) = \frac{1}{2}\big(x + g(x)\big)$ and show that $\omega(x)$ satisfies the conditions of Exercise 5. The result follows since $x = \omega(x)$ if and only if $x = g(x)$.)

7. This problem gives four functions that globally illustrate the four possibilities in the iteration method.

   a) $f(x) = -\frac{1}{2}x - \frac{1}{4}\cos x + \pi$; $g(x) = \frac{1}{2}x - \frac{1}{4}\cos x + \pi$.

   b) $f(x) = -\frac{3}{2}x - \frac{1}{4}\cos x + \pi$; $g(x) = -\frac{1}{2}x - \frac{1}{4}\cos x + \pi$.

   c) $f(x) = x - \frac{1}{4}\cos x + \pi$; $g(x) = 2x - \frac{1}{4}\cos x + \pi$.

   d) $f(x) = -3x - \frac{1}{4}\cos x + \pi$; $g(x) = -2x - \frac{1}{4}\cos x + \pi$.

   Decide which is which and sketch the appropriate picture.

8. Sketch four functions $g_1(x)$, $g_2(x)$, $g_3(x)$ and $g_4(x)$ which satisfy $g(3) = 3$ and $0 \le g_1'(x) < \frac{3}{4}$, $-\frac{3}{4} \le g_2'(x) \le 0$, $g_3'(x) > \frac{5}{4}$ and $g_4'(x) < -\frac{5}{4}$ (respectively) for all real x. Suppose $x_1 = 4$, use $x_{n+1} = g(x_n)$ to sketch the values x_1, x_2, x_3, x_4 and x_5 for each g.

9. Do a complete analysis to find the negative root of $\sqrt{5}$ using Newton's method with $f(x) = x^2 - 5$. Show you obtain quadratic convergence and sketch an appropriate picture.

10. Show that if we apply Newton's method to find the zero of

$$f(x) = x - \frac{3}{4}\sin x - \frac{2}{3},$$

then we get convergence for any starting value to a unique root.

11. In this exercise we will compute the reciprocal of $a > 0$ without using division.

    a) Apply Newton's method to the function $f(x) = \frac{1}{x} - \frac{1}{a}$ to derive the recursion sequence

    $$x_{k+1} = x_k(2 - ax_k).$$

    b) Show that if the initial value x_1 satisfies

    $$0 < x_1 < \frac{2}{a},$$

    then this sequence always converges to $\frac{1}{a}$. (Hint: Try to write a closed form expression for x_n.)

12. Show that if $f(x)$ has a root r of order $m > 1$, i.e., $f(r) = f'(r) = \cdots = f^{(m-1)}(r) = 0$, $f^{(m)}(r) \neq 0$, then Newton's method applied to finding the root $x = r_1$ converges linearly. (Hint: Let $f(x) = (x - r)^m h(x)$ where $h(r) \neq 0$ and expand as in (6a).)

13. Show that if $f(x)$ has a root of order $m > 1$ at $x = r$ then a modified Newton's method with $g(x) = x - \frac{mf(x)}{f'(x)}$ has quadratic convergence in a neighborhood of $x = r$.

14. Let $f(x) = a_3x^3 + a_2x^2 + a_1x + a_0$ be a cubic polynomial with three real roots w_1, w_2, w_3. Show that if Newton's method is started with

$$x_1 = \frac{w_1 + w_2}{2}$$

then the sequence converges to w_3 in one step. (This result is due to H. Walser.)

15. Consider the equation $x^2 - c = 0$, where $c > 0$. Show that if $x_1 > 0$, then the sequence $\{x_k\}$ generated by the Newton-Raphson algorithm always converges. (Hint: Let $y_n = \frac{x_n - \sqrt{c}}{x_n + \sqrt{c}}$ and find a relationship between y_{n+1} and y_n.)

16. Extend the result of Exercise 15 to the general quadratic equation $ax^2 + bx + c = 0$. Distinguish the three cases: no real roots, one double real root, and two distinct real roots.

17. a) Show that if α is a root of order m of the function $f(x)$, then α is a simple root of the function

$$h(x) = \frac{f(x)}{f'(x)}.$$

    b) Prove that the following modification of Newton's method converges quadratically to the roots of $f(x)$ no matter what their multiplicities, providing we start close enough:

$$x_{n+1} = x_n - \frac{f(x_n)f'(x_n)}{f'(x_n)^2 - f(x_n)f''(x_n)}.$$

18. We define the order of an iteration scheme $x_{n+1} = g(x_n)$ to be of order m, where m is a positive number, if the following

$$\lim_{n \to +\infty} \frac{|x_n - r|}{|x_{n+1} - r|^m}$$

is a positive constant, where r is the fixed point of g. Thus Newton's method is of order 2, the method of bisection is order 1 (see Exercise 3, above) and it can be shown that the secant method is of order $\frac{1+\sqrt{5}}{2}$. Show that an mth order method yields approximately m times as many correct significant digits at each step. That is, show that if r is the root to which the iteration is converging and x_n is accurate to k significant digits of r, then x_{n+1} is accurate to approximately km significant digits of r.

Computational Exercises

19. How many zeroes does the function $\cos x - \cos 3x$ have? Find them all.

20. Use the method of bisection to find to six place accuracy all the roots of:

    a) $x^3 - 3x + 1$.

    b) $x^3 - 2\sin x$.

    In each case prove that you have all the roots.

21. We might try another version of the iteration method to solve $x - \frac{3}{4}\sin x - \frac{2}{3}$. Let $g(x) = \arcsin\left(\frac{4}{3}x - \frac{8}{9}\right)$. Try various starting values to see if you can get convergence to the root we know exists. Remember that arcsin is defined only on the interval $[-1, 1]$.

22. Let $g_2(x) = \sqrt{x + \frac{15}{4}}$, $g_3(x) = -\sqrt{x + \frac{15}{4}}$, and $g_4(x) = \frac{15/4}{x-1}$. Analyze each of the functions in turn as a means of finding the roots of the quadratic polynomial $f(x) = x^2 - x - \frac{15}{4}$ as in the text for the function $g_1(x) = \frac{1}{2}(x^2 + x - \frac{15}{4})$. That is, determine to which root these functions converge and for what range of starting values.

23. Use the Newton-Raphson method to compute all the roots of

    a) $\frac{x}{3} = \sin x$

    b) $x^2 = \sin x + \pi^2$

    c) $x^2 = \cos x$

    d) $x^3 - 2x - 5 = 0$ (This problem goes back to the time of Wallis, Newton and Raphson, ca. 1690.)

    In each case prove that you have all the roots.

24. Consider the polynomial

$$x^3 - 10x^2 + 22x + 6 = 0\,.$$

Use Newton-Raphson to find its roots. There is a negative root and two positive roots. Try starting value $x = 2$ and explain the result obtained.

25. a) Let $f(x) = \frac{x+1}{1+x^2}$. The unique root is at $x = -1$. Use Newton's method for various starting values and see which values converge and which don't. Explain the behavior you have observed. (Hint: Graphing $f(x)$ may be useful.) Note that even for those values where the Newton iteration diverges, the function values are decreasing to 0. Thus it is not enough just to watch the function values become small.

    b) Try to find an iteration procedure of the form $x = g(x)$, that converges in the region where the Newton method fails. How fast is the convergence?

26. Use the secant method to find all the roots of

    a) $x^3 - 3x + 1$.

    b) $x^3 - \sin x$.

27. Use the secant method to find all three roots of $x^3 + x^2 - 3x = 0$. Which pairs of starting values converge to which roots?

28. The function $f(x) = e^x - 3x^2$ has three real roots. Find them by the bisection method, the method of iteration, Newton's method and the secant method and compare.

29. Use Newton's method and the secant method to find the root of $x - \tan x = 0$ near $x = 99$ (radians).

30. It is clear that the function $f(x) = x(x-1)^5$ has roots at $x = 0$ and $x = 1$. Which of the methods used in this chapter for finding roots finds both of them fastest?

31. Find the roots, to six places, of the polynomials

    a) $x^4 + 2.2x^3 + 0.77x^2 + 2.2x - .23$

    b) $3x^4 - 3x^3 + 2x^2 - 4$

    c) $x^5 - 12x^4 - 293x^3 + 3444x^2 + 20884x - 240240$

    d) $x^5 + 3x^4 - 5x^3 - 15x^2 + 4x + 12$

    e) $x^5 - 15.8x^4 + 63.8x^3 - 10.6x^2 - 112.8x - 21.6$

    f) $x^4 + 2.2x^3 + .83x^2 - 0.418x + 0.0312$

    g) $x^3 + 4x^2 - 10$

32. Find the roots of the polynomial

$$P(x) = x^6 - 10x^5 + 21x^4 - 82x^3 + 91x^2 - 52x + 12\,,$$

using

a) the straight Newton method;

b) the modified Newton method of part (b) of Exercise 17.

(Hint: all the roots are positive.)

33. The following exercise gives an iteration procedure that can be used to find specified roots of a polynomial. Let

$$P(x) = a_n x^n + a_{n-1}x^{n-1} + \cdots + a_1 x + a_0\,,$$

where $a_n \neq 0$ and a_k is real for $k = 0, 1, \ldots, n$. Suppose that $P(x)$ has n real roots r_i that satisfy

$$0 < r_1 < r_2 < \cdots < r_n\,.$$

a) Suppose we consider the iteration scheme

$$x_{k+1} = -\frac{1}{a_n}\left(a_{n-1} + \frac{a_{n-2}}{x_k} + \cdots + \frac{a_1}{x_k^{n-2}} + \frac{a_0}{x_k^{n-1}}\right).$$

Show that if we start with a value of x_1 sufficiently close to r_n, then the sequence $\{x_k\}$ generated by (3.1) converges to r_n.

b) Suppose we consider the iteration scheme

$$x_{k+1} = -\frac{a_0}{a_1 + a_2 x_k + \cdots + a_n x_k^{n-1}}\,.$$

Show that the sequence $\{x_k\}$ defined by this scheme converges to the root r_1, provided we start close enough to r_1.

c) Use the iterations schemes of parts (a) and (b) to find all the roots of

$$P(x) = x^4 - 10x^3 + 35x^2 - 50x + 24\,.$$

d) Modify the iteration schemes of parts (a) and (b) so that they will converge to the largest and smallest root, respectively, in absolute value.

34. Let $f(x) = e^x - x - 1$ and let $x_0 = 0.5$.

a) Use Newton's method, as given by (3.7), to find the root of $f(x) = 0$.

b) Use the modified form of Newton's method given in Exercise 13 to find the root.

c) Use the modified form of Newton's method given in part (b) of Exercise 17, to find the root.

d) Compare the rates of convergence of each iteration scheme. If there are any differences in rate of convergence, explain. If there are no differences, explain why that is so.

4 *Linear Equations*

The purpose of this chapter is to consider the problem of solving the linear equation $Ax = b$ where A is a matrix and x and b are vectors. In Section 4.1, we begin with a brief introduction to the ideas of linear equations and matrices. We give the connection between these two topics, discuss matrix operations and special matrices and consider solutions of linear equations or their equivalent matrix operations. In Section 4.2, we consider the solutions of linear equations by Gaussian eliminination methods. In particular we show that by using elmentary row operations our original matrix can be put into a canonical form where the solution(s) can be immediately obtained. We also use the rank of a matrix to get the number of solutions. It is hoped that the student has already seen most of the material in the first two sections, but it is not essential.

Unfortunately, in the real world, linear equations have large order and have mostly zero coefficients so that elimination methods are impractical. In this case iteration methods are usually used. In Section 4.3, we consider two basic methods and show that an existence result exists for the convergence to the solution. In Section 4.4, we introduce the study of errors for our linear system through the concept of the condition number of A. The fundamental question is how do changes in A and b affect the solution x? This is important as there are often errors in our representations of these quantities. In Section 4.5, we consider the practical problems of pivoting and scaling linear equations.

4.1 Linear Equations and Matrices

In this section we present an introduction to the topics of matrices and linear equations. These ideas occur in a variety of important real world problems. A special plus of this subject is that unlike nonlinear equations most real problems can be completely solved in a straightforward way with a desired and reasonable degree of accuracy. In most cases, when the number of equations is not excessive a Gaussian elimination method works quite well. These are special methods to handle large systems, particularly where the coefficient matrix is sparse (has a preponderance of zeros), but most of these methods are beyond the scope of this text. We begin with a discussion of matrices and their relation with linear equations. We hope the reader is already acquainted with most of this material.

A *matrix* is a rectangular array of numbers. The matrix

$$A = \begin{bmatrix} a_{11} & a_{12} & \cdots & a_{1m} \\ a_{21} & a_{22} & \cdots & a_{2m} \\ \vdots & & & \vdots \\ a_{n1} & a_{n2} & \cdots & a_{nm} \end{bmatrix}$$

is said to have n *rows* and m *columns*. We will write $A_{n\times m}$ to emphasize the size (dimension) of A or $[a_{ij}]$ where $i = 1, 2, \ldots, n$ and $j = 1, 2, \ldots, m$ to emphasize the individual elements or entries a_{ij} in the ith row and jth column. We will usually denote matrices by A, B, C, ... and their respective elements by a_{ij}, b_{pq}, c_{lm}, ...

There are four basic operations which can be defined for matrices. The *sum* $C = A + B$, is defined if and only if A and B have the same dimension. In this case, $c_{ij} = a_{ij} + b_{ij}$ and C has the common dimension. Similarly, $D = A - B$ with $d_{ij} = a_{ij} - b_{ij}$ is the difference of A and B. The *product* $E = AB$, is defined if and only if the number of columns of A is equal to the number of rows of B. In this case $E_{n\times m} = A_{n\times m} B_{r\times m}$ where $e_{ij} = \sum_{k=1}^{r} a_{ij} b_{jk}$, $(i = 1, 2, \ldots, n; j = 1, 2, \ldots, m)$. In general, $AB \neq BA$. The product $F = \lambda A = [\lambda a_{ij}]$ where λ is a scalar is called *scalar multiplication*.

Finally, $A^T = (b_{ij})$ where $b_{ij} = a_{ji}$ is called the *transpose* of A. We note that $(AB)^T = B^T A^T$ where the product AB is defined.

For example, if

$$A = \begin{bmatrix} 0 & 1 & 2 \\ -3 & 1 & 4 \end{bmatrix} \quad B = \begin{bmatrix} 1 & 2 & -1 \\ 6 & \pi & e \end{bmatrix} \text{ and } C = \begin{bmatrix} 1 & 4 \\ -1 & 0 \\ 2 & 1 \end{bmatrix},$$

the sum $A+C$ is not defined nor is the product AB defined. However the sum

$$A + B = \begin{bmatrix} 2+1 & 1+2 & 2-1 \\ -3+6 & 1+\pi & 4+e \end{bmatrix} = \begin{bmatrix} 3 & 3 & 1 \\ 1 & 1+\pi & 4+e \end{bmatrix}$$

is defined while the product of A and C is

$$AC = \begin{bmatrix} 0 & 1 & 2 \\ -3 & 1 & 4 \end{bmatrix} \begin{bmatrix} 1 & 4 \\ -1 & 0 \\ 2 & 1 \end{bmatrix} = \begin{bmatrix} 3 & 2 \\ 4 & -8 \end{bmatrix}.$$

Similarly, $\pi A = \begin{bmatrix} 0 & \pi & 2\pi \\ -3\pi & \pi & 4\pi \end{bmatrix}$ and $B - A = \begin{bmatrix} 1 & 1 & -3 \\ 9 & \pi - 1 & e-4 \end{bmatrix}$.

Finally,

$$C^T A^T = \begin{bmatrix} 1 & -1 & 2 \\ 4 & 0 & 1 \end{bmatrix} \begin{bmatrix} 0 & -3 \\ 1 & 1 \\ 2 & 4 \end{bmatrix} = \begin{bmatrix} 3 & 4 \\ 2 & -8 \end{bmatrix} = (AC)^T.$$

We now summarize the algebra of matrices. Capital letters denote matrices while lower case letters denote scalars. In Theorem 4.1, "=" means that the i, jth element of the matrix on the left size is equal to the i, jth element on the right side.

Theorem 4.1: *Assuming that the sizes of the matrices are such that the indicated operations can be performed, the following rules of matrix arithmetic are valid.*

(a) $A + B = B + A$

(b) $A + (B + C) = (A + B) + C$

(c) $A(BC) = (AB)C$

(d) $A(B + C) = AB + AC$

(e) $(B + C)A = BA + CA$

(f) $A(B - C) = AB - AC$

(g) $(B - C)A = BA - CA$

(h) $a(B + C) = aB + aC$

(i) $a(B - C) = aB - aC$

(j) $(a + b)C = aC + bC$

(k) $(a - b)C = aC - bC$

(l) $(ab)C = a(bC)$

(m) $a(BC) = (aB)C = B(aC)$

We note that matrices of fixed size $n \times m$ form a commutative group under the $+$ operation. Thus, along with (a) and (b) we have $0 = (z_{ij})$ where $z_{ij} = 0$ for $i = 1, \ldots, n$; $j = 1, \ldots, m$ is the *zero matrix* and satisfies $0 + A = A + 0 = A$ while $-A = (-a_{ij})$ is the additive inverse satisfying $A + (-A) = (-A) + A = 0$.

Multiplication of matrices is much more complicated. For square matrices $A_{n \times n}$ there is an *identity matrix* (under multiplication) $I_{n \times n} = [\delta_{ij}]$ such that $AI = IA = A$. The symbol δ_{ij} is called the *Kroneker Delta*: $\delta_{ij} = 1$ while $\delta_{ij} = 0$ if $i \neq j$. If for a given matrix A, there exists a matrix B such that $AB = BA = I$, then B is the *inverse* of A and we write $B = A^{-1}$. Most matrices do not have this property. A necessary condition that A^{-1} exist is that A be *square*, that is, $m = n$, but this is not a sufficient condition. If A has an inverse we say it is *nonsingular*. A is *singular* if it doesn't have an inverse. Finally, we note that $(AB)^{-1} = B^{-1}A^{-1}$ and $(A^T)^{-1} = (A^{-1})^T$ whenever A and B are nonsingular.

With this background we define a *linear system of n equations in m unknowns* by $Ax - b$ or

$$(4.2) \qquad \begin{aligned} a_{11}x_1 + a_{12}x_2 + \cdots + a_{1m}x_m &= b_1 \\ a_{21}x_1 + a_{22}x_2 + \cdots + a_{2m}x_m &= b_2 \\ &\vdots \\ a_{n1}x_1 + a_{n2}x_2 + \cdots + a_{nm}x_m &= b_m \end{aligned}$$

The matrix A is a known $n \times m$ matrix, called the *coefficient* matrix. The n vector or $n \times 1$ matrix b is also known. The *problem* is to find the solutions of $Ax = b$.

The linear systems

$$\text{(a)} \quad \begin{aligned} x_1 + x_2 &= 2 \\ 2x_1 + 2x_2 &= 3\,, \end{aligned} \qquad \text{(b)} \quad \begin{aligned} x_1 + x_2 &= 2 \\ 2x_1 + 3x_2 &= 4 \end{aligned} \qquad \text{(c)} \quad \begin{aligned} x_1 + x_2 &= 2 \\ 2x_1 + 2x_2 &= 4 \end{aligned}$$

illustrate the fact that there may be none, one, or an infinite number of solutions to a problem. For example, it is impossible to have exactly two solutions for $Ax = b$. These examples illustrate the general case.

Theorem 4.2: *The linear system $Ax = b$ has none, one or an infinite number of solutions.*

If there are no solutions, we say that our system is *inconsistent*. If there is at least one solution, our system is *consistent*. If there is only one solution we say the solution is *unique*. The reader should verify using the algebra of matrices that if $x_1 \neq x_2$ are any two solutions of $Ax = b$, then for any real number t the vector $y = tx_1 + (1-t)x_2$ is also a solution.

Problem 4.1: Verify the result in the last sentence. Is Theorem 4.2 now proven?

4.2 Elimination Methods

In this section we present the Gaussian elimination method which is the best method to solve reasonable problems. In subsequent sections we will see that these methods, when combined with pivoting or scaling operations, form the basis of the standard computer methods used to solve general linear equations. We will break our usual rules of exposition to consider the general case when $m \neq n$. In real life, most problems and computer methods require that $m = n$ and that A have an inverse, so that there is a unique solution. However, for little extra effort we can describe the general situation and then easily reduce it to the case when $n = m$.

There are three elementary row operations which will not change the solution. Thus, they change the problem $Ax = b$ to $Cy = d$, where the set of solutions of the first equation is identical to the set of solutions of the second equation. By the *augmented matrix associated with* $Ax = b$, we mean the $n \times (m+1)$ matrix $[A|b]$ whose first m columns are the respective columns of A and whose $m+1$st column is b.

The three *elementary operations* are

1. multiply any row of the matrix by a constant,
2. interchange the order of any two rows of the matrix,
3. add a multiple of the ith row to the jth row and replace the jth row of the matrix with this sum.

As an example to solve the system

$$\begin{aligned} x_2 + 2x_3 &= 4 \\ x_1 - x_2 + 2x_3 &= 5 \\ 3x_1 \qquad + 3x_3 &= 6 \end{aligned}$$

we begin with the augmented matrix

$$[A|b] = \left[\begin{array}{ccc|c} 0 & -1 & 2 & 4 \\ 2 & -2 & 4 & 10 \\ 3 & 0 & 1 & 6 \end{array}\right].$$

To reduce this matrix we proceed as follows

$$\left[\begin{array}{ccc|c} 0 & -1 & 2 & 4 \\ 1 & -1 & 2 & 5 \\ 3 & 0 & 1 & 6 \end{array}\right] \sim (A) \left[\begin{array}{ccc|c} 1 & -1 & 2 & 5 \\ 0 & -1 & 2 & 4 \\ 3 & 0 & 1 & 6 \end{array}\right]$$

$$\sim (B) \left[\begin{array}{ccc|c} 1 & -1 & 2 & 5 \\ 0 & -1 & 2 & 4 \\ 0 & 3 & -5 & -9 \end{array}\right] \sim (C) \left[\begin{array}{ccc|c} 1 & -1 & 2 & 5 \\ 0 & 1 & -2 & -4 \\ 0 & 3 & -5 & -9 \end{array}\right]$$

$$\sim (D) \left[\begin{array}{ccc|c} 1 & -1 & 2 & 5 \\ 0 & 1 & -2 & -4 \\ 0 & 0 & 1 & 3 \end{array}\right] \sim (E) \left[\begin{array}{ccc|c} 1 & -1 & 2 & 5 \\ 0 & 1 & 0 & 2 \\ 0 & 0 & 1 & 3 \end{array}\right]$$

$$\sim (F) \left[\begin{array}{ccc|c} 1 & -1 & 0 & -1 \\ 0 & 1 & 0 & 2 \\ 0 & 0 & 1 & 3 \end{array}\right] \sim (G) \left[\begin{array}{ccc|c} 1 & 0 & 0 & 1 \\ 0 & 1 & 0 & 2 \\ 0 & 0 & 1 & 3 \end{array}\right].$$

Our steps (A)~(G) are in excruciating detail for illustrative purposes. The student will do better combining steps. At each step we have computed an equivalent augmented matrix whose associated system has the solution $x_1 = 1$, $x_2 = 2$, $x_3 = 3$. Note that elementary operation 2 is done in step (A), elementary operation 1 is done in (C), while elementary operation 3 is done in the remaining steps. By *forward elimination* we mean the steps (A) to (D) where our augmented matrix is now in upper triangular form. The augmented matrix

$$\left[\begin{array}{ccc|c} 1 & -1 & 2 & 5 \\ 0 & 1 & -2 & -4 \\ 0 & 0 & 1 & 3 \end{array}\right] \tag{4.3}$$

is usually said to be in *row-reduced form*. The solution is now found by simple calculations: clearly $x_3 = 3$, while $x_2 - 2x_3 = -4$ can be solved to yield $x_2 = 2$ and $x_1 - x_2 + 2x_3 = 5$ can be solved to yield $x_1 = 1$. The process from (E) to (G) is usually called *back-substitution.* If we repeat this example by replacing row 1 with row 3 and then multiplying the new row 1 by $\frac{1}{3}$ we will be lead to an equivalent augmented matrix to (4.3) and the same final matrix in back-substitution.

Problem 4.2: Carry out the steps described in the last sentence.

If one does steps (A)~(D) and back-substitution then the method is usually called Gaussian elimination. If one does the steps (A)~(G) then the method is usually called Gauss-Jordan elimination.

The general method works even if there is not a unique solution. Thus, for the system (a) above we have

$$\left[\begin{array}{cc|c} 1 & 1 & 2 \\ 2 & 2 & 3 \end{array}\right] \sim \left[\begin{array}{cc|c} 1 & 1 & 2 \\ 0 & 0 & 1 \end{array}\right]$$

which demonstrates that there is no solution since $0x_1 + 0x_2 = 1$ is not possible. Similarly, for the system (c) above we have

$$\left[\begin{array}{cc|c} 1 & 1 & 2 \\ 2 & 2 & 4 \end{array}\right] \sim \left[\begin{array}{cc|c} 1 & 2 & 2 \\ 0 & 0 & 0 \end{array}\right]$$

which demonstrates that the second equation was redundant and that there are an infinite number of solutions (x_1, x_2) satisfying $x_1 + x_2 = 2$.

We note that these elementary operations can be simulated or performed by multiplication of special matrices. This is especially important when using a computer or programming subroutines to solve linear algebra problems. In our example, a matrix M is found by performing the corresponding elementary operation on $I_{3\times 3}$. Thus, to perform (A) we interchange the first and second rows of I to get $M_1 = \begin{bmatrix} 0 & 1 & 0 \\ 1 & 0 & 0 \\ 0 & 0 & 1 \end{bmatrix}$, to perform (B) we use $M_2 = \begin{bmatrix} 1 & 0 & 0 \\ 0 & 1 & 0 \\ -3 & 0 & 1 \end{bmatrix}$, to perform (C) we use $M_3 = \begin{bmatrix} 1 & 0 & 0 \\ 0 & -1 & 0 \\ 0 & 0 & 1 \end{bmatrix}$, etc. Now the product

$$M_3M_2M_1[A|b] = M_3M_2 \begin{bmatrix} 0 & 1 & 0 \\ 1 & 0 & 0 \\ 0 & 0 & 1 \end{bmatrix} \left[\begin{array}{ccc|c} 0 & -1 & 2 & 4 \\ 1 & -1 & 2 & 5 \\ 3 & 0 & 1 & 6 \end{array}\right]$$

$$= M_3 \begin{bmatrix} 1 & 0 & 0 \\ 0 & 1 & 0 \\ -3 & 0 & 1 \end{bmatrix} \left[\begin{array}{ccc|c} 1 & -1 & 2 & 5 \\ 0 & -1 & 2 & 4 \\ 3 & 0 & 1 & 6 \end{array}\right]$$

$$= \begin{bmatrix} 1 & 0 & 0 \\ 0 & -1 & 0 \\ 0 & 0 & 1 \end{bmatrix} \left[\begin{array}{ccc|c} 1 & -1 & 2 & 5 \\ 0 & -1 & 2 & 4 \\ 0 & 3 & -5 & -9 \end{array}\right]$$

$$= \left[\begin{array}{ccc|c} 1 & -1 & 2 & 5 \\ 0 & 1 & -2 & -4 \\ 0 & 3 & -5 & -9 \end{array}\right].$$

The reader should continue this process to simulate the complete elimination (A)~(G) above.

Problem 4.3: Carry out the steps suggested in the last sentence.

We now consider the threory of solutions for problems in linear algebra using elementary operations for general n and m. The case when $m = n$ is an immediate consequence of our discussion. We assume we have used forward elimination and back-substitution so that $[A|b] \sim E = [C|d]$ where E is *reduced row-echelon form*, that is,

1. the rows of E consisting entirely of zeros are grouped together as the bottom rows of E.
2. if a row has a nonzero entry, its first nonzero entry is one (called the leading one of the row).
3. nonzero rows of E are in order of their leading ones.
4. each column with a leading one has zero entries everywhere else in the column.

The following matrices are in reduced row-echelon form:

$$\begin{bmatrix} 0 & 0 \\ 0 & 0 \end{bmatrix}, \quad \begin{bmatrix} 1 & 0 & 0 \\ 0 & 1 & 0 \\ 0 & 0 & 1 \end{bmatrix}, \quad \begin{bmatrix} 1 & -3 & 0 \\ 0 & 0 & 1 \\ 0 & 0 & 0 \end{bmatrix} \quad \text{and} \quad \begin{bmatrix} 1 & 0 & 0 & 0 \\ 0 & 0 & 1 & 0 \\ 0 & 0 & 0 & 1 \end{bmatrix}$$

while the following matrices are not:

$$\begin{bmatrix} 0 & 0 \\ 1 & 0 \end{bmatrix}, \quad \begin{bmatrix} 0 & 1 \\ 1 & 0 \end{bmatrix}, \quad \begin{bmatrix} 1 & 0 & 1 \\ 0 & 1 & 0 \\ 0 & 0 & 1 \end{bmatrix} \quad \text{and} \quad \begin{bmatrix} 2 & 0 \\ 0 & 1 \end{bmatrix}.$$

To consider solutions of $Ax = b$ we need three integers n, m, and k and a picture of an augmented system which has been put into reduced row-echelon form $[C|d]$. As above, n is the number of equations and m is the number of unknowns. The integer k is the number of leading ones of $[C]$. It is usually called the *dimension* or *rank* of A and is defined to be the largest integer k so that A has a $k \times k$ submatrix which has an inverse or nonzero determinant. For those students of linear algebra k is also the number of linearly independent rows or columns of A.

What do we make of a system which has been reduced to

$$[C|d] = \left[\begin{array}{cccc|c} 1 & z_1 & 0 & 0 & z_2 \\ 0 & 0 & 1 & 0 & z_3 \\ 0 & 0 & 0 & 1 & z_4 \\ 0 & 0 & 0 & 0 & y_1 \\ 0 & 0 & 0 & 0 & y_2 \end{array}\right] ?$$

In this example $n = 5$, $m = 4$, and $k = 3$. Our original problem had five equations in four unknowns, that is, A is 5×4, b is 5×1 and the unknown vector x is 4×1. The z characters in $[C|d]$ can be any values, but the y characters determine if there will be a solution.

Our first observation is that there is at least one solution if and only if y_1 and y_2 are each zero. Thus, if y_1 is not zero there is no solution of $Ax = b$. If y_1 and y_2 are zero, our original problem had two redundant equations; that is there are $n - k = 2$ rows of zeros in reduced form. There is also an $m - k = 1$ parameter family of solutions. That is, if we parameterize one unknown (for example set $x_2 = t$), there is a unique solution for the remaining unknowns. In this example where z_1, z_2, z_3, and z_4 are real numbers, $x_4 = z_4$, $x_3 = z_3$, $x_2 = t$ and $x_1 = z_2 - t$. In vector notation we have

$$x = \begin{bmatrix} x_1 \\ x_2 \\ x_3 \\ x_4 \end{bmatrix} = \begin{bmatrix} z_2 \\ 0 \\ z_3 \\ z_4 \end{bmatrix} + t \begin{bmatrix} -1 \\ 1 \\ 0 \\ 0 \end{bmatrix}$$

which is a line in four space. There can not be a unique solution since $n > k$.

If $b \equiv 0$ our problem is said to be *homogeneous*. In this case there is at least one solution, namely $x \equiv 0$. Our example illustrates the general theorem, that the solution of $Ax = b$ is of the form $x_n + x_p$ where x_p is a particular solution and x_n is any solution of $Ax = 0$.

If $n = m = k$ then A is invertible with inverse A^{-1} and $Ax = b$ has a unique solution given by $x = A^{-1}b$. For a variety of problems it is necessary to find the inverse of an $n \times n$ matrix A. The usual procedure is to reduce the augmented matrix $[A|I_{n\times n}]$ to $[I|B]$. If this can be done then $B = A^{-1}$. For example, to find the inverse of $A = \begin{bmatrix} 2 & 3 \\ 1 & 2 \end{bmatrix}$ we have

$$\left[\begin{array}{cc|cc} 2 & 3 & 1 & 0 \\ 1 & 2 & 0 & 1 \end{array}\right] \sim \left[\begin{array}{cc|cc} 1 & 2 & 0 & 1 \\ 2 & 3 & 1 & 0 \end{array}\right] \sim \left[\begin{array}{cc|cc} 1 & 2 & 0 & 1 \\ 0 & -1 & 1 & -2 \end{array}\right]$$

$$\sim \left[\begin{array}{cc|cc} 1 & 2 & 0 & 1 \\ 0 & 1 & -1 & 2 \end{array}\right] \sim \left[\begin{array}{cc|cc} 1 & 0 & 2 & -3 \\ 0 & 1 & -1 & 2 \end{array}\right].$$

Problem 4.4: The reduction in the last sentence can be thought of as a simultaneous reduction of $Ax = b_1$ and $Ax = b_2$. Use this to describe the columns of A^{-1}.

Problem 4.5: Make up an example of A where $n = m = 2$ and $k = 1$ and use the reduction to find A^{-1}. What can you conclude about this method for finding A^{-1}? Repeat this exercise when $m = k = 2$ and $m = 3$.

As an example we consider the case when A is as above and $b = [8, 5]^T$. In this case we have the unique solution

$$x = A^{-1}b = \begin{bmatrix} 2 & -3 \\ -1 & 2 \end{bmatrix} \begin{bmatrix} 8 \\ 5 \end{bmatrix} = \begin{bmatrix} 1 \\ 2 \end{bmatrix}.$$

4.3 Iteration Techniques

As opposed to direct methods such as Gaussian elimination methods, it is often preferable to use an iterative methods. Iteration methods are used when the matrix A is sparse (many zeros) without a specific pattern of zeros or when self correcting methods are useful such as in performing hand calculations. The *purpose* of this section is to introduce the reader to this subject by considering the two most popular methods: Jacobi's method and the Gauss-Seidel method. In this section, A is an $n \times n$ matrix and b is an n-vector.

We define an *iteration technique* for solving $Ax = b$ to be a formula

$$x^{(k+1)} = Tx^{(k)} + c \tag{4.4}$$

where the $n \times n$ matrix T and the $n \times 1$ vector c are chosen so that $x = Tx + c$ is consistent with $Ax = b$. The notation $x^{(k)}$ is used for the kth iterate of the sequence$\{x^{(k)}\}$. The first element of this sequence (initial guess) is denoted by $x^{(1)}$. We say the sequence $\{x^{(k)}\}$ *converges* to the vector $x^{(0)}$, and write $x^{(k)} \to x^{(0)}$ if, for each $j = 1, \ldots, n$, the jth component $\{x_j^{(k)}\}$ is convergent to the jth component of x^0; that is, $x_j^{(k)} \to x_j^{(0)}$.

To get T for our iteration methods using matrices we set $A = D - L - U$ where

$$D = \begin{bmatrix} a_{11} & & & \\ & a_{12} & & 0 \\ & & \ddots & \\ 0 & & & a_{1n} \end{bmatrix}, \tag{4.5a}$$

$$L = -\begin{bmatrix} 0 & & & & \\ a_{21} & 0 & & & \\ \vdots & & \ddots & & \\ \vdots & & & \ddots & \\ a_{n1} & a_{n2} & \cdots & a_{n,n-1} & 0 \end{bmatrix} \tag{4.5b}$$

and

$$U = -\begin{bmatrix} 0 & a_{12} & & \cdots & a_{1n} \\ & 0 & a_{23} & \cdots & a_{2n} \\ & & \ddots & & \vdots \\ & & & \ddots & a_{n-1,n} \\ & & & & 0 \end{bmatrix} \tag{4.5c}$$

If $a_{ii} \neq 0$ for each $i = 1, \ldots, n$ then D^{-1} is diagonal and invertible. Thus, $b = Ax = Dx - (L + U)x$ implies the iteration method

$$x^{(k+1)} = D^{-1}(L + U)x^{(k)} + D^{-1}b \tag{4.6}$$

is consistent with $Ax = b$ if $T = D^{-1}(L + U)$ and $c = D^{-1}b$.

If we denote the jth component of $x^{(k)}$ by $x_j^{(k)}$ then the matrix equation (4.6) becomes

$$x_i^{(k+1)} = \frac{\left[-\sum_{j=1, j\neq i}^{n} a_{ij} x_j^{(k)} + b_i \right]}{a_{ii}}, \quad i = 1, \ldots n. \tag{4.7}$$

This is called the *Jacobi iteration method.* If we use the latest values of $x_j^{(k)}$ in (4.6) we have the *Gauss-Seidel method* which is written as

$$x_i^{(k+1)} = \frac{\left[-\sum_{j=1}^{k-1} a_{ij}x_j^{(k+1)} - \sum_{j=i+1}^{n} a_{ij}x_j^{(k)} + b_i\right]}{a_{ii}}, \tag{4.8}$$

$i = 1, \ldots, n$.

We note that although (4.7) and (4.8) may appear formidable to the reader, they are in a simple, convenient form to be used in a computer program to compute $x_k^{(k+1)}$. It is expected that Gauss-Seidel usually converges faster than Jacobi's method since we are using updated results, but there are examples of A and b where one method converges but the other does not.

These methods are easier to understand if we use an example. Thus if our problem is

$$\begin{aligned} 3x_1 + x_2 - x_3 &= 2 \\ -x_1 + 4x_2 + x_3 &= 10 \ , \\ -x_1 + x_2 - 3x_3 &= -8 \end{aligned}$$

the Jacobi iteration method is

$$\begin{aligned} x_1^{(k+1)} &= \frac{-x_2^{(k)} + x_3^{(k)} + 2}{3}, \\ x_2^{(k+1)} &= \frac{x_1^{(k)} - x_3^{(k)} + 10}{4}, \\ \text{and}\quad x_3^{(k+1)} &= \frac{x_1^{(k)} - x_2^{(k)} - 8}{-3}, \end{aligned}$$

while the Gauss-Seidel method is

$$x_1^{(k+1)} = \frac{-x_2^{(k)} + x_3^{(k)} + 2}{3},$$

$$x_2^{(k+1)} = \frac{x_1^{(k+1)} - x_3^{(k)} + 10}{4},$$

$$\text{and} \quad x_3^{(k+1)} = \frac{x_1^{(k+1)} - x_2^{(k+1)} - 8}{-3}.$$

In Tables 4.1 and 4.2 we give the iteration values for each method starting with $x_1^{(1)} = 0$, $x_2^{(1)} = 0$, $x_3^{(1)} = 0$.

In general, we might want to modify schemes (4.7) and (4.8) so that when we solve for the variable x_i, we take the remaining equation that has the largest coefficient in absolute value for x_i. For example, if we were given

$$\begin{aligned} 2x_1 + 3x_2 - x_3 &= 7 \\ x_1 + 5x_2 + 2x_3 &= -1 \\ 3x_1 - x_2 + 3x_3 &= 2 \end{aligned}$$

we would choose

$$\begin{aligned} x_1 &= \frac{x_2 - 3x_3 + 2}{3} && \text{(from the third equation)} \\ x_2 &= \frac{-x_1 - 2x_3 - 1}{5} && \text{(from the second equation)} \\ x_3 &= 2x_1 + 3x_2 - 7 && \text{(from the first equation)} \end{aligned}$$

rather than the straightforward choice

$$x_1 = \frac{-3x_2 + x_3 + 7}{2}$$

$$x_2 = \frac{-x_1 - 2x_3 - 1}{5}$$

$$x_3 = \frac{-3x_1 + x_2 + 2}{3}.$$

n	x_1	x_2	x_3
1	0	0	0
2	0.6666666666	2.5	2.66666666
3	0.72222222	2	3.27777778
4	1.09259259	1.86111111	3.09259259
5	1.07716049	2	2.9228395
6	0.9742798333	2.03858024	2.97427983
7	0.97856653	2	3.02143347
8	1.00714449	1.98928326	3.00714449
9	1.00595374	2	2.99404625
10	0.9980154166	2.00297687	2.99801542
11	0.9983461833	1.99999999	3.00165382
12	1.00055127	1.99917309	3.00055127
13	1.00045939	2	2.9995406
14	0.9998468666	2.0022969	2.99984687
15	0.9998723933	1.9999999	3.00012761
16	1.00004254	1.9993619	3.00004253
17	1.00003544	2	2.99996455
18	0.9999881833	2.00001772	2.99998818
19	0.9999901533	2	3.00000984
20	1.00000328	1.99999507	3.00000328

Table 4.1: Jacobi's Iteration Scheme

The reasons for this modification are to attempt to make the system diagonally dominant (see Theorem 4.3 below) and are related to the concept of pivoting which will be discussed below in Section 4.5.

This example illustrates a very important result which holds if A if *strictly diagonally dominant*, that is, if $|a_{ii}| > \sum_{j=1, j\neq i}^{n} |a_{ij}|$ holds for $i = 1, 2, \ldots, n$. We note that many problems can be put into this form by interchanging the rows of A. The reader should observe that this result is similar to the result in Chapter 3 where $x = g(x)$ and $|g'(x)| \leq k < 1$.

n	x_1	x_2	x_3
1	0	0	0
2	0.6666666666	2.66666666	3.33333333
3	0.88888889	1.88888889	3
4	1.03703703	2.00925925	2.99074074
5	0.9938271633	2.0007716	3.00231481
6	1.0005144	1.99954989	2.99967849
7	1.00004286	2.00009109	3.00001607
8	0.9999749933	1.99998973	3.00000491
9	1.00000506	2.00000003	2.99999832
10	0.99999943	2.00000027	3.00000028
11	1	1.99999993	2.99999997
12	1.00000001	2.00000001	3
13	0.9999999966	1.99999999	3
14	1	2	3
15	1	2	3

Table 4.2: Gauss-Seidel Iteration Scheme

Theorem 4.3: *If A is strictly diagonally dominant, then for any choice of $x^{(1)}$, the sequence $\{x^{(k)}\}$ generated by either the Jacobi method or the Gauss-Seidel method converges to the solution $x^{(0)}$ of $Ax = b$.*

Finally, we recall that in Chapter 3 we gave results for the rate of convergence of iteration processes. Similar results can be obtained for linear iteration techniques. This is done in Theorem 4.4 of Section 4.4 after we understand more about vector and matrix norms.

4.4 Matrix Norms and Condition Number

The problem we often face in the numerical solution of linear equations is to decide how a small change in b or A changes our solution x of $Ax = b$. The result is best stated using the condition number of A and will be done in Theorem 4.4 of this section. At the same time we must introduce an important generalization of lengths or distances, called norms. This concept is a fundamental part of applied mathematics and will be used to give bounds on the errors of solutions.

Intuitively, by the norm of the an element or vector x, we mean a nonnegative real number $||x||$ which measures the length of x. If x and y are given, we use the symbolism $||x - y||$ to denote their distance. We will temporarily defer the usual mathematical definitions and instead define a specific norm for an $n \times 1$ vector x and an $n \times n$ matrix A. The right hand side of (4.10), below, is usually called the *maximum row-sum.*

Definition 4.1: *If $x = [x_1, \ldots, x_n]^T$ and $A = [a_{ij}]$, we define*

$$||x||_\infty = \max_{1 \le i \le n} |x_i| \tag{4.9}$$

and

$$||A||_\infty = \max_{1 \le i \le n} \sum_{j=1}^{n} |a_{ij}| \,. \tag{4.10}$$

We have chosen (4.9) and (4.10) so as to preserve an important property of norms of linear transformation, which is

$$||A|| = \max_{||x||=1} ||Ax|| = \max_{||x|| \ne 0} \frac{||Ax||}{||x||} \,. \tag{4.11}$$

The last equality follows since $||ax|| = |a|\,||x||$ and $||A(ax)|| = |a|\,||Ax||$ for any real number a.

Thus,

$$||Ax|| \le ||A||\,||x|| \,. \tag{4.12}$$

Problem 4.6: The result in (4.12) suggests that $||A||$ satisfies a minimum property. State and justify this property.

For example, if $A = \begin{bmatrix} 1 & -1 & 2 \\ 0 & 1 & -1 \\ 2 & 1 & 0 \end{bmatrix}$ and $x = \begin{bmatrix} 1 \\ 0 \\ -2 \end{bmatrix}$ then

$$\sum_{j=1}^{3} |a_{1j}| = |1| + |-1| + |2| = 4\,,$$

$$\sum_{j=1}^{3} |a_{2j}| = |0| + |1| + |-1| = 2 \qquad \text{and}$$

$$\sum_{j=1}^{3} |a_{3j}| = |2| + |1| + |0| = 3$$

so that $||A||_\infty = 4$. Note that $y = Ax = [-3, 2, 2]^T$, $||x||_\infty = 2$, $||y||_\infty = 3$ so that $3 = ||y||_\infty = ||Ax||_\infty < ||A||_\infty \, ||x||_\infty = 4 \cdot 3 = 12$.

The idea of norm or length occurs in many areas of the mathematical sciences. The idea is that we have a vector space $(V, +, \cdot)$ where V is the set of elements closed under the addition operation "+" and the scalar operation "$\cdot$". Then $|| \; || : V \to \mathbb{R}$ is a *norm* if for x, y in V and α in $\mathbb{R}$ we have

(i) $||x|| \geq 0, \quad ||x|| = 0$ if and only if $x = 0$,

(ii) $||\alpha x|| = |\alpha| \, ||x||$, and

(iii) $||x + y|| \leq ||x|| + ||y||$.

It is important to observe that in Definition 4.1 and the above discussion (i)–(iii) hold for the vectors x in $\mathbb{R}^n$ and A in the set $M_{n\times n}$ of $n \times n$ matrices. In the special case of $M_{n\times n}$ we require a further condition

(iv) A, B in $M_{n\times n}$ implies $||AB|| < ||A|| \, ||B||$

and refer to $||A||$ as a *matrix norm.*

We note that there are other important examples of norms of vectors x and matrices A in addition to $||x||_\infty$ and $||A||_\infty$ discussed above. We refer the reader to Burden [6] for this material which is beyond the scope of our objectives.

For most vectors we have the inequality $||Ax|| < ||A||\,||x||$ which relates the length of $y = Ax$ to the length of x. For a relatively few vectors, the equality $||Ax|| = ||A||\,||x||$ is obtained. If y is an approximation to x and the nonnegative number $||x - y||$ represents the error in approximating x by y, the resulting error in finding Ay as opposed to Ax satisfies

$$||Ax - Ay|| = ||A(x - y)|| \leq ||A||\,||x - y||\,. \tag{4.13}$$

That is, error is magnified by no more than $||A||$. The result follows from (f) of Theorem 4.3, that is, $A(x - y) = Ax - Ay$ for any two vectors x and y.

We are now ready to give the result promised in the last section which is found in Burden [6]. A more general result is also found in the same reference. This theorem, which deals with rates of convergence, logically belongs directly after Theorem 4.3 of Section 4.3, but was postponed until we had discussed the concept of norms.

Theorem 4.4: *If $||T|| < 1$ in any matrix norm then the sequence $\{x^{(k)}\}$, generated using (4.6), converges to the unique solution x for any choice of the initial value $x^{(1)}$. The following estimates for the error bound hold:*

$$||x - x^{(k)}|| \leq ||T||^{k-1}||x^{(1)} - x|| \qquad \textit{and}$$

$$||x - x^{(k)}|| \leq \frac{||T||^{k-1}}{1 - ||T||}\,||x^{(2)} - x^{(1)}||\,.$$

We now return to our basic question which is how will the error in b affect the error in x for the linear system $Ax = b$. We remind the reader that A is invertible since we assumed there is a unique solution $x = A^{-1}b$. Thus, the nonnegative number

$$\text{cond}(A) = ||A||\,||A^{-1}|| \tag{4.14}$$

is well defined. This number is called the *condition number* of A. Note that the condition number is at least 1 since $1 = ||I|| = ||AA^{-1}|| \leq ||A||\,||A^{-1}||$. The reader is cautioned that generally $||A^{-1}|| \neq ||A||^{-1}$ (see Exercise 4.2). For an approximate solution $\overline{x}$, we also defined the *error*, $e = x - \overline{x}$ and the *residual* $r = b - A\overline{x}$.

The point is that $\overline{x}$ and hence r are known to us after our calculations while x and hence e are unknown. Our specific problem is whether we can estimate the values of the unknowns x and e by using the values of the knowns, $\overline{x}$ and r. The answer is that this can be done using several applications of the inequality (4.12). We first note that $r = b - A\overline{x} = Ax - A\overline{x} = A(x - \overline{x}) = Ae$ and $e = A^{-1}r$ so that $||r|| \leq ||A||\,||e||$ and $||e|| \leq ||A^{-1}||\,||r||$. Thus,

$$\frac{||r||}{||A||} \leq ||e|| \leq ||A^{-1}||\,||r||\,. \tag{4.15}$$

Since $Ax = b$ and $x = A^{-1}b$, we also have $||b|| \leq ||A||\,||x||$ and $||x|| \leq ||A^{-1}||\,||b||$. Combining these inequalities with (4.15) we have

$$\frac{1}{\mathrm{cond}A}\frac{||r||}{||b||} \leq \frac{||e||}{||x||} \leq \mathrm{cond}A\frac{||r||}{||b||}\,. \tag{4.16}$$

Theorem 4.5: *For a given approximate solution $\overline{x}$ of $Ax = b$ with residual $r = b - A\overline{x}$, the error $||e|| = ||x - \overline{x}||$ and the relative error $\frac{||e||}{||x||}$ satisfy* (4.15) *and* (4.16) *respectively provided that $||b|| \neq 0$.*

In Exercise 4.1 we have asked the reader to give a careful proof of Theorem 4.5 while in Exercises 4.2 and 4.3 we ask the reader to verify (4.15) and (4.16) with a specific example. Finally, we note that the above calculations lead to an alternate method of solution for this problem.

Unfortunately, there are times when even the best algorithms will not lead to an acceptable solution of $Ax = b$. This may occur when the condition number is large. These are systems whose matrices of coefficients are ill-conditioned, that is, the solution is extremely sensative to round-off errors. The only hope is to try more precision and be careful in your calculations. In general, the larger the system the greater the precision required since there is more opportunity for round-off error.

As an example, if

$$A = \begin{bmatrix} 3.02 & -1.05 & 2.53 \\ 4.33 & 0.56 & -1.78 \\ -0.83 & -0.54 & 1.47 \end{bmatrix}$$

and

$$b = \begin{bmatrix} -1.61 \\ 7.23 \\ -3.38 \end{bmatrix},$$

we have, after Gaussian elimination with pivoting and carrying three significant digits,

$$\begin{bmatrix} 4.33 & 0.56 & -1.78 & 7.23 \\ 0 & -1.44 & 3.77 & -6.65 \\ 0 & 0 & -0.00362 & 0.00962 \end{bmatrix}.$$

This gives the calculated solution $\overline{x} = \begin{bmatrix} 0.880 \\ -2.35 \\ -2.60 \end{bmatrix}$, whereas the actual solution is

$$x = \begin{bmatrix} 1 \\ 2 \\ -1 \end{bmatrix}.$$

If we carry six figures, then we get the improved calculated solution

$$\overline{x} = \begin{bmatrix} 0.9998 \\ 1.9995 \\ -1.002 \end{bmatrix}.$$

Using three significant digits, the residual is

$$r = b - A\overline{x} = \begin{bmatrix} 0.0053 \\ 0.0008 \\ -0.0084 \end{bmatrix}$$

which is a smaller error than the error $e = x - \overline{x}$. Using (4.14), the condition number is

$$\text{cond}A = ||A||\,||A^{-1}|| = (6.67)(1118.7) = 7595\,,$$

which is relatively large. Using (4.16), the relative error $\frac{||e||}{||x||}$ satisfies

$$1.5297 \times 10^{-7} \leq \frac{||e||}{||x||} \leq 8.8241$$

so that the relative error could be quite large. Using (4.15), we also have

$$1.2594 \times 10^{-3} \leq ||e|| \leq 9.5651\,.$$

Thus, the error could be quite large.

It is important to note that because the condition number of the matrix of coefficients is so large our estimates of the relative error and the error are very poor. This situation is due to the fact that A is ill-conditioned and is not significantly improved by more precision in our calculations. In fact, in our example,

$$||e|| = 4.35 \quad \text{and} \quad ||x|| = 2\,,$$

so that

$$\frac{||e||}{||x||} = 2.175\,.$$

If the condition number is near 1, then the residual gives a very good estimate of the error made in calculating the solution x since our equality (4.15) would look roughly like

$$\frac{1}{1+\epsilon}\frac{||r||}{||b||} \leq \frac{||e||}{||x||} \leq (1+\epsilon)\frac{||r||}{||b||},$$

where ϵ is some small positive constant.

If the condition number is large, then we must increase our precision to compensate for this problem, or if this is not possible, consider the use of an iteration method to finish the problem. For example, if $\frac{||r||}{||b||} \approx 10^{-8}$, then $\frac{1}{7595}10^{-8} \leq \frac{||e||}{||x||} \leq (7595)10^{-8} \approx 10^{-4}$ so that our relative error is correct to approximately four places. Thus, if our computer has eight place accuracy we can not guarantee a relative error good to five places.

Finally we note that iteration methods and elimination methods may be combined. We use elimination methods to obtain starting values for the iteration methods which in some sense are self-correcting.

4.5 Pivoting and Scaling

In this section we consider some practical problems of solving $Ax = b$. We consider the ideas of pivoting and scaling since they are indispensible in solving real world problems.

Pivoting is the operation of putting the maximum elements, in absolute value, along the main diagonal of the $n \times n$ matrix A. This can be accomplished by interchanging rows and columns. However, in solving a system of equations one does not usually wish to interchange columns since this also interchanges variables and in a large system this could cause too much time to be given over to mere bookkeeping. Instead, we do what is called partial pivoting, that is, we only interchange rows as we work our way down the main diagonal using Gaussian elimination. This amounts to only interchanging some equations in the system of equations and the bookkeeping is quite easy.

If we use pivoting in combination with Gaussian elimination, we help ensure that we have nonzero elements in the main diagonal. If we cannot get any nonzero numbers into the diagonal spots we are presently working on, we find that our matrix is singular. Pivoting also helps to increase the accuracy of our computations. To illustrate this aspect of pivoting consider the system of equations $Ax = b$, where

$$A = \begin{bmatrix} -0.002 & 4.000 & 4.000 \\ -2.000 & 2.906 & -5.387 \\ 3.000 & -4.031 & -3.112 \end{bmatrix}$$

and

$$b = \begin{bmatrix} 7.998 \\ -4.481 \\ -4.143 \end{bmatrix} .$$

If we simply do the Gaussian elimination without partial pivoting, then we end up with the following upper triangular augmented matrix

$$\left[\begin{array}{ccc|c} -0.002 & 4.000 & 4.000 & 7.998 \\ 0.000 & -3.997 & -4.005 & -8.003 \\ 0.000 & 0.000 & -10.00 & 0.000 \end{array}\right] ,$$

which gives us the computed solution

$$x = \begin{bmatrix} -1496 \\ 2.000 \\ 0.000 \end{bmatrix} .$$

Since the exact solution is

$$x_1 = 2\,, \quad x_2 = 0\,, \quad \text{and} \quad x_3 = 1\,,$$

we see that round-off error has caught up with us, in particular in the small value of a_{11} compared to all the other entries. If we pivot first, that is, interchange rows one and two, we get the augmented matrices

$$\left[\begin{array}{ccc|c} 3.000 & -4.031 & -3.112 & -4.413 \\ -0.002 & 4.000 & 4.000 & 7.998 \\ -2.000 & 2.906 & -5.387 & -4.481 \end{array}\right]$$

and

$$\left[\begin{array}{ccc|c} 3.000 & -4.031 & -3.112 & -4.413 \\ 0.000 & 3.997 & 3.998 & 7.995 \\ 0.000 & 0.000 & -7.681 & -7.681 \end{array}\right].$$

whose computed solution is

$$x_1 = 2, \quad x_2 = 0, \quad \text{and} \quad x_3 = 1.$$

Here the error is zero.

Scaling is the process of dividing each entry in a row by the largest entry, in absolute value, in that row. This is quite beneficial if the entries in a row are disparate in magnitude, but if the entries all have roughly the same magnitude, all that scaling will do is to produce more round-off error from the extra divisions.

We give an example to illustrate the benefits of scaling. Consider the system of equations $Ax = b$, where

$$A = \begin{bmatrix} 1 & -1 & 100 \\ 2 & 1 & 100 \\ -1 & 2 & 1 \end{bmatrix}$$

and

$$b = \begin{bmatrix} 100 \\ 103 \\ 2 \end{bmatrix},$$

whose solution is clearly

$$x_1 = 1, \quad x_2 = 1, \quad \text{and} \quad x_3 = 1.$$

If we go through the process of Gaussian elimination with partial pivoting, but without scaling, then we end up with the augmented matrix

$$\left[\begin{array}{ccc|c} 1 & 0.5 & 50 & 32.3 \\ 0 & 1.0 & -33.3 & -32.3 \\ 0 & 0.0 & 134.3 & 133.3 \end{array}\right],$$

whose computed solution is

$$x_1 = 0.99\,, \quad x_2 = 0.67\,, \quad \text{and} \quad x_3 = 1.67\,,$$

which is significantly different than the correct solution. If we do scaling first, we would divide the first row of the augmented matrix by 100, the second row by 100 and the third row we would leave alone since all the entries are of the same magnitude. If we now do Gaussian elimination with partial pivoting, then we end up with the augmented matrix

$$\left[\begin{array}{ccc|c} 1 & -2 & -1 & -2 \\ 0 & 1 & 101 & 102 \\ 0 & 0 & -4.03 & -4.03 \end{array}\right].$$

whose computed solution is

$$x_1 = 1\,, \quad x_2 = 1\,, \quad \text{and} \quad x_3 = 1\,,$$

which is the correct solution.

Exercise Set 4

1. Derive (4.15) carefully.

2. If $A = \begin{bmatrix} 1 & 2 \\ 0 & 1 \end{bmatrix}$ compute A^{-1} and show that cond$A = 9$.

3. Let $b = \begin{bmatrix} 5 \\ 2 \end{bmatrix}$, $A = \begin{bmatrix} 1 & 2 \\ 0 & 1 \end{bmatrix}$, and $\overline{x} = \begin{bmatrix} 1.05 \\ 1.95 \end{bmatrix}$. Verify the inequalities (4.14) and (4.15).

4. Show that the condition number of a nonsingular matrix A is at least equal to 1.

5. a) Show that if one uses Gaussian elimination and back-substitution to solve an $n \times n$ system, then it would take at most

$$\frac{n(n-1)(2n-1)}{6} + n(n-1) + \frac{n(n+1)}{2},$$

multiplication and divisions.
(Hint: $1+2+\cdots+n = \frac{1}{2}n(n+1)$ and $1^2+2^2+\cdots+n^2 = \frac{1}{6}n(n+1)(2n+1)$.)

   b) Give an upper bound for the total number of operations needed to solve the same $n \times n$ system using the Gauss-Jordan method.

6. Prove that if x_1 and x_2 are solutions to the equation $Ax = b$, then for any real number t the vector $y = tx_1 + (1-t)x_2$ is also a solution to the equation $Ax = b$.

7. Solve the system of equations

$$\begin{aligned} 15x_1 - 2x_2 - 6x_3 \phantom{{}-21x_4} &= 300 \\ -2x_1 + 12x_2 - 4x_3 - x_4 &= 0 \\ -6x_1 - 4x_2 + 19x_3 - 9x_4 &= 0 \\ -x_2 - 9x_3 + 21x_4 &= 0\,. \end{aligned}$$

This system arises in studying the current through a closed electrical system.

8. a) Find the inverse of the matrix

$$\begin{bmatrix} 3 & 2 & -5 \\ 2 & -3 & 1 \\ 1 & 4 & -1 \end{bmatrix}.$$

b) Use this fact to find the solution to the system

$$\begin{aligned} 3x_1 + 2x_2 - 5x_3 &= 1 \\ 2x_1 - 3x_2 + x_3 &= -1 \\ x_1 + 4x_2 - x_3 &= 0\,. \end{aligned}$$

9. a) Solve the system $Ax = b$, where

$$A = \begin{bmatrix} -0.002 & 4.000 & 4.000 \\ -2.000 & 2.906 & -5.387 \\ 3.000 & -4.031 & -3.112 \end{bmatrix}$$

and

$$b = \begin{bmatrix} 7.998 \\ -4.481 \\ -4.143 \end{bmatrix}$$

by using Gaussian elimination without partial pivoting. Is the solution you found really a solution to the system? You are only allowed to keep 4 significant figures. Do not scale, i.e. do not put 1's on the main diagonal.

b) Add partial pivoting to your routine and solve. Is this solution any better than the solution found in part (a)? Again, do not scale.

10. Solve the system

$$\begin{aligned} x_1 + \tfrac{1}{2}x_2 + \tfrac{1}{3}x_2 + \tfrac{1}{4}x_4 &= 1 \\ \tfrac{1}{2}x_1 + \tfrac{1}{3}x_2 + \tfrac{1}{4}x_3 + \tfrac{1}{5}x_4 &= -1 \\ \tfrac{1}{3}x_1 + \tfrac{1}{4}x_2 + \tfrac{1}{5}x_3 + \tfrac{1}{6}x_4 &= 1 \\ \tfrac{1}{4}x_1 + \tfrac{1}{5}x_2 + \tfrac{1}{6}x_3 + \tfrac{1}{7}x_4 &= 0\,. \end{aligned}$$

Will the computation cause any problem for the computer? What is the condition number for the matrix of coefficients?

11. Let E be the error made in inputing the matrix A so that we are really solving the system $(A+E)x = b$. Let $\overline{A} = A + E$ and let $\overline{x}$ be the solution to $\overline{Ax} = b$. Show that

$$\frac{||x - \overline{x}||}{||\overline{x}||} \leq \text{cond}(A)\frac{||E||}{||A||}.$$

12. Solve the system

$$\begin{bmatrix} 3 & 2 & -1 & -4 \\ 1 & -1 & 3 & -1 \\ 2 & 1 & -3 & 0 \\ 0 & -1 & 8 & -5 \end{bmatrix} \begin{bmatrix} x_1 \\ x_2 \\ x_3 \\ x_4 \end{bmatrix} = \begin{bmatrix} 2 \\ 3 \\ 1 \\ 3 \end{bmatrix}.$$

13. Solve the system

$$\begin{aligned} 2.51x_1 + \ 1.48x_2 + \ 4.53x_3 &= 0.05 \\ 1.48x_1 + \ 0.93x_2 - \ 1.30x_3 &= 1.03 \\ 2.68x_1 + \ 3.04x_2 - \ 1.48x_3 &= -0.53 \end{aligned}$$

a) by Gaussian elimination without pivoting;

b) by Gaussian elimination with pivoting.

In each case carry just three significant digits and chop off. Note which gives you the better solution. Redo part (a) but this time carry six significant. How does this effect your answers?

14. Solve Exercise 13 by using the Gauss-Seidel iterative method.

15. Solve the 2×2 system

$$\begin{aligned} ax_1 + \quad x_2 &= 1 \\ x_1 + \quad x_2 &= 2 \end{aligned}$$

without pivoting and with pivoting. In both cases do not do any algebraic simplification, e.g., leave $\frac{1}{a}$ as $\frac{1}{a}$. If a is a very small quantity (say 10^{-8}), what are the appropriate solutions? Is it better to pivot or not?

16. a) Without rearranging the equations to make them diagonally dominant use both the Jacobi and Gauss-Seidel iterative schemes to solve the system

$$\begin{aligned} 1.1x_1 + \quad 2x_2 + \quad 6.1x_3 &= -5.12 \\ -3.1x_1 + \quad 1.1x_2 - \quad x_3 &= -1.36 \\ 0.3x_1 + \quad 2.2x_2 + \quad 0.9x_3 &= -0.06\,. \end{aligned}$$

Does either method converge?

b) Run the part (a) again, but this time rearrange the equations to make them diagonally dominant. Does either method converge this time?

5 *Interpolation and Curve Fitting*

Let S denote the set of $n+1$ points $(x_0, y_0), (x_1, y_1), \ldots (x_n, y_n)$ where $x_0 < x_1 \cdots < x_n$. The purpose of this chapter is to consider the problem of finding a function $f(x)$, such as a polynomial or a piecewise polynomial (spline), such that the graph of f contains the $n+1$ points of S. If this happy event occurs (as in Figure 5.1 below), we say that f *interpolates* S. The ideas of this chapter are often associated with the problems of interpolation and curve fitting.

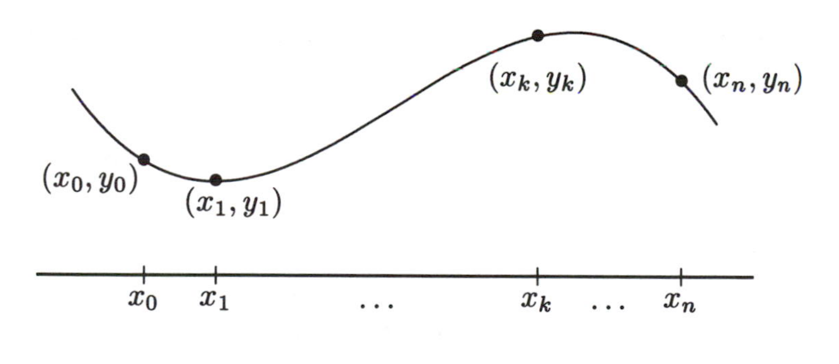

Figure 5.1

We begin with the subject of polynomial interpolation. The reasons that we often use interpolation by polynomials are many: they are easy to evaluate, pleasing to the eye, infinitely differentiable, etc. The main result is that there is a unique polynomial of degree n or less which interpolates the $n+1$ points of S. We will also see how this polynomial is constructed and determine the interpolation error if the points of S come from a smooth function f. Unfortunately, in many cases, polynomials and their derivatives oscillate too much to be of practical value. In Section 5.2 we will consider interpolation by piecewise polynomials or splines and in particular piecewise polynomials of degree one and three.

Section 5.3 deals with curve fitting of the set S in a least squares sense by a straight line. These results immediately generalize to the problem of fitting S in a least squares sense by an arbitrary function f on $[a,b]$. These problems are very rich in mathematical theory and are beyond the scope of this book.

5.1 Interpolation by Polynomials

We begin this section with two theorems. The first is an important theoretical result but not useful in any practical sense. The second theorem is of both theoretical and practical importance.

Theorem 5.1: (Weierstrass Approximation Theorem) *If f is a continuous function on $[a,b]$ then for any $\epsilon > 0$ there exists a polynomial $p(x)$ such that*

$$\max_{x \text{ in } [a,b]} |f(x) - p(x)| < \epsilon \,. \tag{5.1}$$

We have pictured this result in Figure 5.2, below. What the theorem says is that for any $\epsilon > 0$ (no matter how small) we can find a polynomial whose graph is between the graph of the functions $y = f(x) - \epsilon$ and $y = f(x) + \epsilon$ respectively. We sometimes say that f is "uniformly approximated" by $p(x)$ on $[a,b]$ in that it holds for all x in $[a,b]$. It is very important for us to know that such a polynomial exists. However, in general the degree of the required

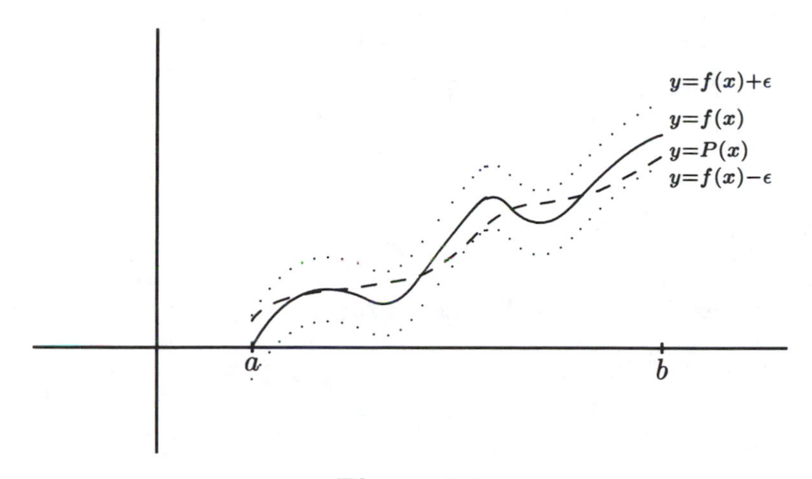

Figure 5.2

polynomial is so large as to make this result impractical to use. The interested reader will find further details in Bartle [3]. Details of this construction are given in Exercise 5.23.

The next theorem is the main theorem on polynomial interpolation. We remind the reader that $S = \{(x_i, y_i) : i = 0, 1, 2, \ldots, n\}$. We will soon construct the interpolating polynomial, but for several reasons we will first sketch the proof of this theorem.

Theorem 5.2: *Through any set S of $n+1$ distinct points, there exists a unique polynomial of degree n or less which interpolates S.*

The existence of this unique polynomial follows from our knowledge of elementary linear algebra. We begin by setting $y = p(x) = A_0 + A_1 x + \cdots + A_n x^n$. The problem is to determine the coefficients $A_0, A_1, \ldots, A_n$ satisfying $y_i = p(x_i)$ for $i = 0, 1, \ldots, n$. Thus, we have the matrix equation $VA = Y$ where

$$V = \begin{bmatrix} 1 & x_0 & x_0^2 & \cdots & x_0^n \\ 1 & x_1 & x_1^2 & \cdots & x_1^n \\ & \vdots & & & \\ 1 & x_n & x_n^2 & \cdots & x_n^n \end{bmatrix}, \quad A = \begin{bmatrix} A_0 \\ A_1 \\ \vdots \\ A_n \end{bmatrix} \text{ and } Y = \begin{bmatrix} y_0 \\ y_1 \\ \vdots \\ y_n \end{bmatrix}.$$

The real number $\det V$ is called the *Vandermonde determinant.* Its value is given by

$$\det V = \prod_{\substack{i,j=0 \\ i>j}}^{n} (x_i - x_j) = (x_1 - x_0)(x_2 - x_1)(x_2 - x_0) \cdots$$
$$(x_n - x_{n-1})(x_n - x_{n-2}) \cdots (x_n - x_0)$$

which is never zero since $x_i \neq x_j$ if $i \neq j$. Thus, from linear algebra, V is invertible and there is a unique solution $A = V^{-1}Y$.

Problem 5.1: Find the equation of the line containing the points (3,2) and (-1,-2) using the results in the last two paragraphs.

For completeness, we note the possibility that $A_n = 0$ so that our polynomial has degree strictly less than n. For example, if the set S contains collinear points then $A_2 = A_3 = \cdots = A_n = 0$.

Problem 5.2: Find the polynomial of degree two or less which interpolates the points (-1,2), (0,3) and (1,4).

There is an elementary proof of Theorem 5.2 using the Lagrange Interpolation Formula given in (5.2) and (5.3) (below). Using this formula we will construct a polynomial $p(x)$ of degree n or less such that $y_i = p(x_i)$ for $i = 0, 1, \ldots, n$. To show this polynomial is unique we note that if there is a second polynomial $g(x)$ with these properties then $h(x) = p(x) - g(x)$ is a polynomial of degree n or less which vanishes at $x = x_0, x_1, \ldots, x_n$. By elementary algebra $h(x)$ is either the zero polynomial or has degree greater than $n+1$. Thus, $h(x) \equiv 0$ for all x or $p(x) = g(x)$ so that the interpolating polynomial is unique.

Our next task is to construct the interpolating polynomial which interpolates S. There are many methods of doing this, but we will consider the easiest and most illustrative method which is called Lagrange interpolation. The best method of explanation is to "jump right in," therefore, let

$$\ell_j(x) = \frac{(x-x_0)(x-x_1)\cdots(x-x_{j-1})(x-x_{j+1})\cdots(x-x_n)}{(x_j-x_0)(x_j-x_1)\cdots(x_j-x_{j-1})(x_j-x_{j+1})\cdots(x_j-x_n)}, \quad j = 0, 1, 2, \ldots, n. \tag{5.2}$$

We note that each $\ell_j(x)$ is a polynomial of degree n and that $\ell_j(x_i) = \delta_{ij}$ where $\delta_{ii} = 1$ and $\delta_{ij} = 0$ if $i \neq j$. Thus,

$$P(x) = \sum_{j=0}^{n} y_j \ell_j(x) \tag{5.3}$$

is a polynomial of degree less than or equal to n with

$$P(x_i) = \sum_{j=0}^{n} y_j \delta_{ij} = y_i \,.$$

By Theorem 5.2, this must be the unique interpolating polynomial for S.

For example, if $S = \{(x_0, y_0), (x_1, y_1)\}$ then

$$\ell_0(x) = \frac{(x - x_1)}{(x_0 - x_1)} \quad \text{and} \quad \ell_1(x) = \frac{(x - x_0)}{(x_1 - x_0)}$$

so that

$$\begin{aligned}
P(x) &= y_0\ell_0(x) + y_1\ell_1(x) \\
&= y_0\frac{x - x_1}{x_0 - x_1} + y_1\frac{x - x_0}{x_1 - x_0} \\
&= x\left(\frac{y_1 - y_0}{x_1 - x_0}\right) + \frac{x_1y_0 - x_0y_1}{x_1 - x_0}.
\end{aligned}$$

The reader should verify that $P(x) = mx + b$ where m is the slope and b is the y-intercept of the straight line containing (interpolating) the points (x_0, y_0) and (x_1, y_1).

For another example of Lagrange interpolation we will find a quadratic interpolating polynomial for $\sin x$ based on the points $x =$ 0, 0.5 and 1.0. Thus we want our polynomial, $P(x)$, to go through the points $(0, \sin 0)$, $(0.5, \sin 0.5)$ and $(1, \sin 1)$. (See Figure 5.3, below). The Lagrange formula gives

$$\begin{aligned}
P(x) = \sin(0)&\frac{(x - 0.5)(x - 1.0)}{(0 - 0.5)(0 - 1.0)} + \sin(0.5)\frac{(x - 0)(x - 1.0)}{(0.5 - 0)(0.5 - 1.0)} \\
&+ \sin(1.0)\frac{(x - 0)(x - 0.5)}{(1.0 - 0)(1.0 - 0.5)} \\
&= x\big[2(\sin 1.0 - 2\sin 0.5)x + 4\sin 0.5 - \sin 1\big].
\end{aligned}$$

In Exerercises 5.1–5.5 we have included a variety of illustrative problems. In particular, we show that even though there is a unique interpolating polynomial for S of degree n or less, it may not give good answers in terms of smoothness and fit.

As we have emphasized several times, assuming we have an interpolating polynomial, we would like some idea of the error involved. The answer depends upon where the data comes from. In the next theorem we assume that there exists a function f defined on the interval $[a, b]$ so that the $(n + 1)$st derivative, $f^{(n+1)}(x)$, is continuous on $[a, b]$. We also assume that $y_i = f(x_i)$ for each point (x_i, y_i) in S.

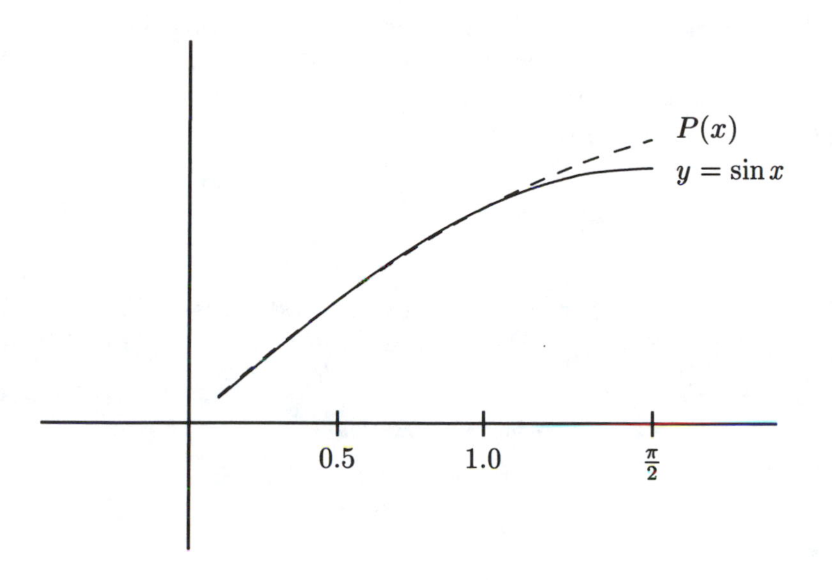

Figure 5.3

Theorem 5.3: *If f is as above, $x_0, x_1, \ldots, x_n$ are distinct points in $[a, b]$ and $P(x)$ is the unique interpolating polynomial satisfying $y_i = f(x_i)$, $i = 0, 1, \ldots, n$ then for any x in $[a, b]$, there exists $\xi = \xi(x)$ in (a, b) such that*

$$(5.4)\quad E(x) = f(x) - p(x) = \frac{f^{(n+1)}(\xi)}{(n+1)!}(x - x_0)(x - x_1)\cdots(x - x_n)\,.$$

The proof is given in Exercise 5.7. It involves multiple applications of Rolle's Theorem and we encourage the reader to work through the details.

Problem 5.3: Let $f(x) = \sin x$ and $P(x)$ be as on the previous page. Find ξ in Theorem 5.3 when $x = 0.75$.

To understand some of the major ideas of interpolation and error with a minimum of effort we consider examples where $f(x) = mx+b$. If $S : (x_0, y_0)$ then by Theorem 5.2, the interpolating polynomial is $P_0(x) = y_0$ and the error is

$$E(x) = f'(\xi)(x - x_0) = m(x - x_0)$$

which implies that $f(x) = m(x - x_0) + y_0$ which is correct since it is the unique $1st$ degree polynomial with slope m which contains the point (x_0, y_0).

Similarly, if $S : (x_0, y_0), (x_1, y_1)$ then by Theorem 5.2 $p(x) = mx + b$ and

$$E(x) = \frac{f''(\xi)}{2!}(x - x_0)(x - x_1) = 0$$

which is correct. Finally, if S contains $n > 2$ collinear points, $f^{(n+1)}(\xi) = 0$ as expected.

We also illustrate Theorem 5.3 with an estimate of the error in calculating sin(0.25) using our quadratic interpolating polynomial that we obtained above. We have

$$\begin{aligned} \sin 0.25 &= (0.5)\big[2(\sin 1.0 - 2\sin 0.5)(0.25) + 4\sin 0.5 - \sin 1.0\big] \\ &\approx 0.25438528 . \end{aligned}$$

Our error estimate above says that

$$\begin{aligned} E(0.25) &= (0.25 - 0)(0.25 - 0.5)(0.25 - 1.0)\frac{-\cos\xi}{3!} \\ &= 0.0078125(-\cos\xi) , \end{aligned}$$

where ξ is in the interval $[0, 1.0]$, since $\cos x$ is decreasing on the interval $[0, 1.0]$ we can obtain upper and lower bounds for the error. We have

$$(0.0078125)(-\cos 0) \le E(0.25) \le (0.0078125)(-\cos 1)$$

or

$$-0.0078125 \leq E \leq -0.0042212\,.$$

Since the error is negative, we see that our estimated value is too large. Indeed

$$\sin 0.25 \approx .247403959$$

so that the error made is really $-.00698132$ to this accuracy.

By way of comparison we shall compute the error made from the cubic Taylor polynomial approximation to $\sin x$ about $x = 0$, namely

$$Q(x) = x - \frac{x^3}{6}\,.$$

We have

$$Q(0.25) = (0.25) - \frac{(0.25)^3}{6} = 0.247395833\,,$$

so that our error is -8.12625×10^{-6}. We might expect a better result since we are using a cubic polynomial for our interpolation. Taylor's Theorem tells us

$$\sin x = x + \frac{f'''(\xi)x^3}{6}\,,$$

where ξ is some point between 0 and x. If we use this result we obtain for the error estimate the value, 0.00260417. Thus, in either case the Taylor polynomial gives the better approximation at $x = 0.25$.

In spite of the comparison above, interpolation results are rather impressive. Our Taylor series answers would have gotten worse if we had chosen x closer to 1. Most importantly, in the majority of cases we do not know the function $f(x)$ associated with S or can not easily give a Taylor series expansion.

Problem 5.4: Find the errors in using $P(x)$ and $Q(x)$ when $x = 0.75$. Do these errors agree with the theoretical results such as the bounds on E given above when $x = 0.25$?

5.2 Interpolation by Piecewise Polynomials

We have seen in Section 5.1 that we can solve the problem of interpolating $n+1$ points. Unfortunately, polynomial interpolation is not a practical method of interpolation because polynomials oscillate too much. An example such as the one pictured below in Figure 5.4 should easily convince the reader that the four points P_1, P_2, P_3, P_4 from the curve $y = f(x)$ are very poorly interpolated by the third order polynomial $p_3(x)$.

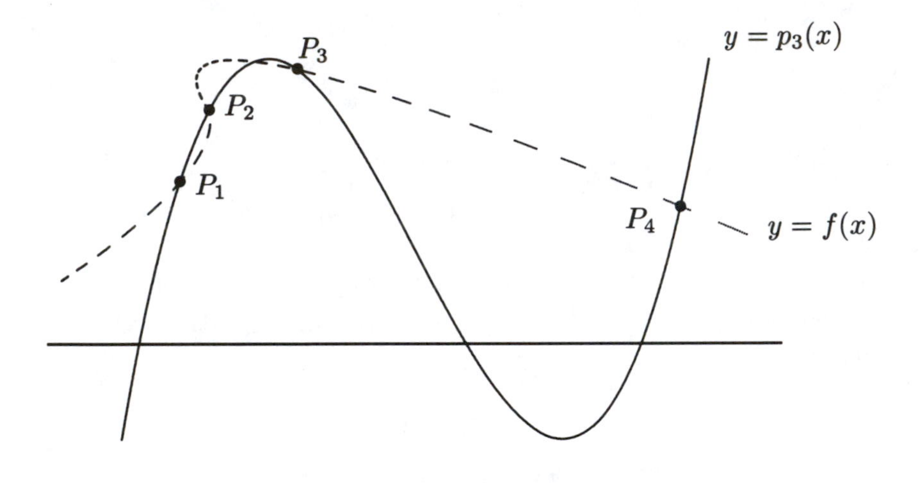

Figure 5.4

This problem is very important. When the first author was much younger he worked for a company which designed ramjets. The task was to simulate the flow of a gaseous mixture. In Figure 5.5 we have pictured the flow of this mixture in our container and in Figure 5.6 we have pictured the calamity which can occur.

In Figure 5.5, below, we picture the flow influenced by the smooth boundary $y = f(x)$. Everything is nice and we can build a mathematical model which agrees with wind tunnel experiments

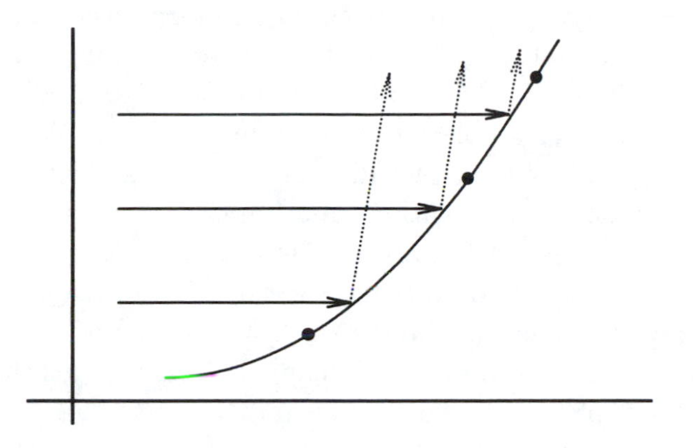

Figure 5.5

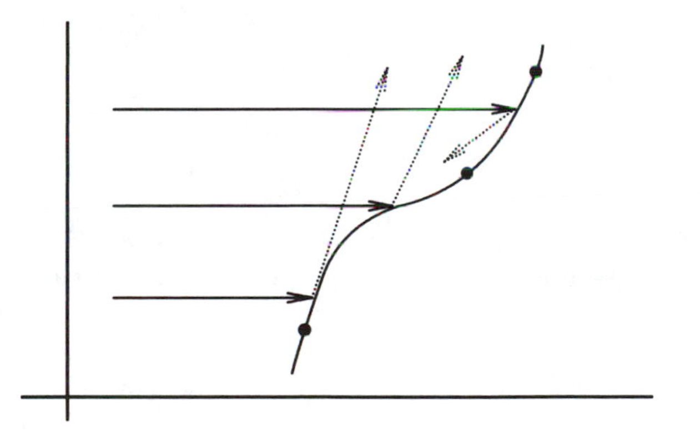

Figure 5.6

with amazing accuracy. In Figure 5.6, above, everything is as before except that we have interpolated the three points P_1, P_2 and P_3. This interpolation causes total chaos in that it introduces shock waves into our model where none exist. This is an impossible situation and must be corrected.

The modern way to provide smooth interpolation is by use of piecewise polynomials or splines. If the reader carefully examines the problems in Figures 5.4, 5.5 and 5.6 or in Exercises 5.1 and 5.2 she might note that the problem with polynomial interpolation is that the values of $p'(x)$ or its higher derivatives are too large in comparison with a smooth interpolating function that we would choose. In Exercise 5.15a we see that splines satisfy a minimizing property for the derivatives of the interpolating function. For example, among all interpolating functions $g(x)$ for the set $S = \{(x_i, y_i) : i = 0, 1, \ldots, n\}$, $x_0 < x_1 < \cdots < x_n$ the piecewise linear function $S_1(x)$ is the solution to the problem of finding the minimum of $I_1(g)$ where

$$I_1(g) = \int_{x_0}^{x_n} g'^2(x)\, dx\,.$$

(See Exercise 5.15a.) Similarly, a piecewise cubic polynomial interpolating function $S_3(x)$ gives the minimum of $I_3(g)$ where

$$I_3(g) = \int_{x_0}^{x_n} g''^2(x)\, dx\,.$$

(See Exercise 5.15b.) We picture in Figure 5.7, below, the situation for the piecewise linear case with $n = 4$ and remind the reader that

$$\int_{x_0}^{x_4} S''^2_3(x)\, dx \le \int_{x_0}^{x_4} g''^2(x)\, dx$$

for any interpolating function $g(x)$.

Perhaps, a brief history is in order. Splines are a modern success story! They were introduced by Schoenberg in 1946. The name "splines" refers to a tool used by draftsmen for smooth interpolation. It is a flexible rod with weights which pass through the points to be interpolated in a smooth manner from one interval to the next. Since 1960, the technical literature of splines has increased at an exponential rate. The first author "discovered" splines in his Ph.D. thesis in 1969 where their minimizing properties were of interest.

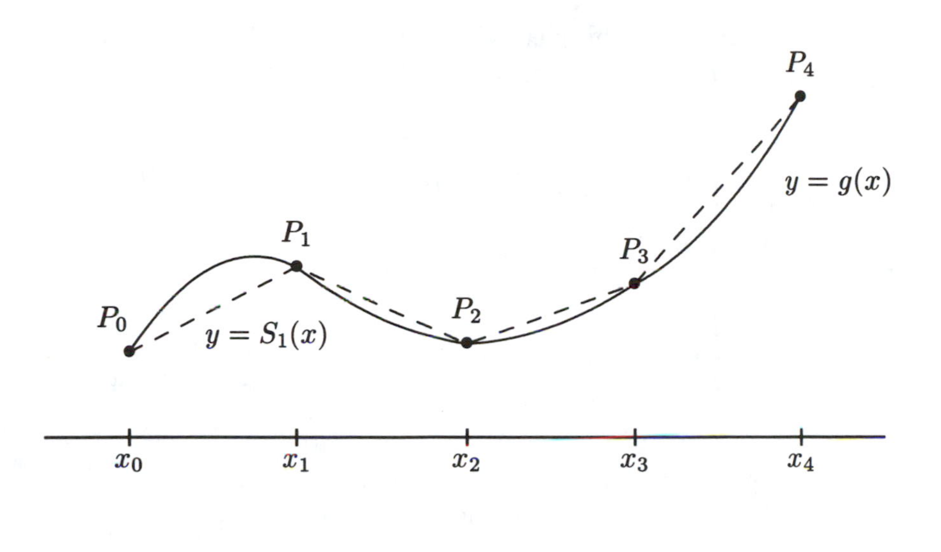

Figure 5.7

Practically, they are the most widely used interpolating tool. In particular, cubic splines are most often used since they provide a very pleasing, smooth fit. We know, for example, that two dimensional cubic splines are used to design automotive fenders because of their smooth fit.

We will consider piecewise linear splines and piecewise cubic splines. Linear splines have a tremendous advantage in that it is easy to interpolate with them. Unfortunately, if $s(x)$ is such a function, $s'(x)$ is not continuous which is a major disadvantage. Cubic splines often offer the best of both worlds. It is not too difficult to construct them and they are extremely smooth (their second derivative is continuous).

By a *spline of degree one* (piecewise linear) we mean a function $s(x)$ continuous on $[a, b]$ such that if $a = x_0 < x_1 < \cdots < x_n = b$ is a partition of $[a, b]$ then $s(x)$ is linear on each subinterval $[x_i, x_{i+1}]$, $i = 0, 1, \ldots, n-1$. The points x_k are called *knots.*

The interpolation by piecewise linear splines is easy. Thus, if $\{(x_0, y_0), \ldots, (x_n, y_n)\}$ is a set of data points then

$$(5.5) \quad s(x) = \begin{cases} y_0 + \dfrac{x - x_0}{x_1 - x_0}(y_1 - y_0), & x_0 \le x \le x_1 \\ y_1 + \dfrac{x - x_1}{x_2 - x_1}(y_2 - y_1), & x_1 < x \le x_2 \\ \vdots & \\ y_{n-1} + \dfrac{x - x_{n-1}}{x_n - x_{n-1}}(y_n - y_{n-1}), & x_{n-1} < x < x_n . \end{cases}$$

We also note that s is a linear combination of basis functions $z_0(x), z_1(x), \ldots, z_n(x)$ pictured below in Figure 5.8 for $n = 4$.

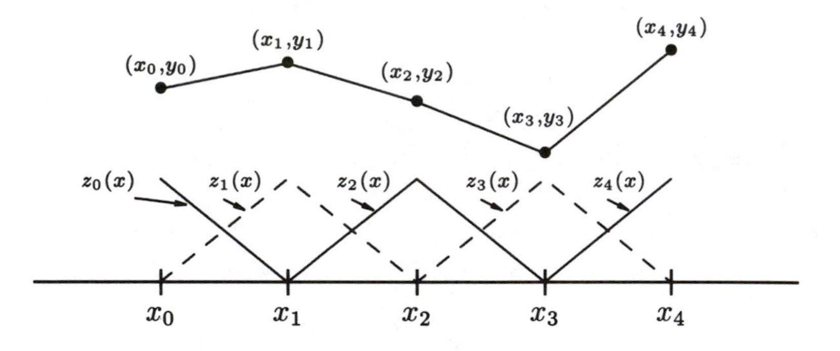

Figure 5.8

Thus, if $z_k(x)$ is such that $z_k(x_k) = 1$, $z_k(x_\ell) = 0$ for $k \ne \ell$ and $z_k(x)$ is linear between knots then

$$(5.6) \qquad s(x) = \sum_{k=0}^{n} y_k z_k(x) .$$

We give an example of linear spline interpolation. Let $f(x) = x^3 + 2x^2 - x + 1$. Consider the points $(-2,3)$, $(-1,3)$, $(0,1)$, $(1,3)$, and $(2,15)$ which lie on the graph of $y = f(x)$. By (5.5), we see that the linear spline that interpolates these points is given by

$$S(x) = \begin{cases} 3 & -2 \le x \le -1 \\ 3 - 2(x+1) & -1 \le x \le 0 \\ 1 + 2x & 0 \le x \le 1 \\ 3 + 12(x-1) & 1 \le x \le 2\,. \end{cases}$$

Note that $f(0.5) = 1.125$ and $S(0.5) = 2$. This discrepancy is due to the fact that the curve has a local minimum at $(-2+\sqrt{7})/3 \approx 0.215$ so that $S(x)$ lies significantly above the curve. This reflects the fact that linear splines are not smooth curves.

Problem 5.5: Sketch a picture of $f(x)$ and $S(x)$ for this last example.

Problem 5.6: Show that the inequality in Exercise 5.15a holds for the last example.

We now turn to cubic splines. Unlike the linear splines considered above, cubic splines are not only continuous, but their first and second derivatives are continuous. These additional continuity properties are enough to make cubic splines quite useful in fitting data sets or interpolating functions since they imply that the curvature of our spline is continuous. We shall see that cubic splines are not as easy to construct as linear splines, or as easy to write down, but, nevertheless, because of the smoothness they provide, they are often well worth the extra effort.

We now give the general definition of a cubic spline.

Definition 5.1: *Let f be a function defined on the interval $[x_1, x_n]$ and let $x_1 < x_2 < \cdots < x_n$. Suppose that for $j = 1, 2, \ldots, n$ we have $f(x_j) = y_j$. A* **cubic spline interpolant**, *S for the function f passes through the n given points is a function that satisfies the following conditions.*

a) *S is a cubic polynomial, denoted by S_j, in each subinterval $[x_j, x_{j+1}]$, for $j = 1, \ldots, n-1$.*

b) *For each $j = 1, \ldots, n$ we have*

$$S(x_j) = y_j \,.$$

c) *For each $j = 1, \ldots, n-2$ we have*
   (i) $S_{j+1}(x_{j+1}) = S_j(x_{j+1})$,
   (ii) $S'_{j+1}(x_{j+1}) = S'_j(x_{j+1})$
   and
   (iii) $S''_{j+1}(x_{j+1}) = S''_j(x_{j+1})$.

We note that condition (c) says that S, S' and S'' are continuous.

d) *Furthermore, the end values of $S''(x)$ must be specified in some way, that is, we must specify two conditions that $S''(x_1)$ and $S''(x_n)$ must satisfy.*

   The most common conditions on $S''(x_1)$ and $S''(x_n)$ are the following
   (iv) $S''_1(x_1) = S''_n(x_n) = 0$,
   (v) $S''_1(x_1) = S''_2(x_2)$ *and* $S''_n(x_n) = S''_{n-1}(x_{n-1})$
   (vi) $S''_1(x_1)$ *and* $S''_n(x_n)$ *are linear interpolations of the two nearest values of* $S''(x)$

If the cubic spline satisfies the condition (iv), then $S(x)$ is called a *natural spline*. This is the most often used form of cubic spline.

If the cubic spline satisfies the condition (vi) and if the data is actually fit by a cubic polynomial, then the spline is this polynomial. Neither of the other two conditions possess this property. The condition (vi) is equivalent to assuming that

$$S'(x_1) = f'(x_1) \quad \text{and} \quad S'(x_n) = f'(x_n) \,.$$

In order to write the condition (vi) in a form which can be used for calculations, we introduce some notation that will be useful in what follows. For $j = 1, \dots, n-1$ we define

$$h_j = x_{j+1} - x_j$$

and for $j = 1, \dots, n$ we define

$$G_j = S_j''(x_j)\,.$$

Thus by (vi), we can write

$$G_1 = \frac{(h_1 + h_2)G_2 - h_1 G_3}{h_2}$$

and

$$G_n = \frac{(h_{n-2} + h_{n-1})G_{n-1} - h_{n-1}G_{n-2}}{h_{n-2}}\,.$$

One can prove that there is a unique cubic spline interplant that satisfies (a), (b), (c) and any one of (iv), (v), or (vi). Indeed, our construction that follows will be a sketch of such a proof.

If we write $S_j(x)$ as follows

$$S_j(x) = a_j(x - x_j)^3 + b_j(x - x_j)^2 + c_j(x - x_j) + d_j\,, \tag{5.7}$$

then we must determine a_j, b_j, c_j, and d_j so that $S_j(x)$ satisfies the conditions of the definition above for a cubic spline.

By (b) of the definition, we know that $S_j(x_j) = y_j$ and so, by (5.7), we have

$$d_j = y_j\,.$$

Note that from (5.7)

$$S_j''(x) = 6a_j(x - x_j) + 2b_j$$

and so we have, by (vi) of the definition

$$G_j = 2b_j \quad \text{and} \quad G_{j+1} = 6a_j h_j + 2b_j$$

or, solving for a_j and b_j, we have

$$b_j = \frac{G_j}{2} \quad \text{and} \quad a_j = \frac{G_{j+1} - G_j}{6h_j}.$$

By (iv) of the definition we must have

$$y_{j+1} = ajh_j^3 + b_j h_j^2 + c_j h_j + d_j$$

and so, if we use the values obtained for a_j, b_j and d_j, we find that

$$c_j = \frac{y_{j+1} - y_j}{h_j} - h_j \frac{2G_j + G_{j+1}}{6}.$$

Finally, if we calculate $S_j'(x)$ and apply condition (v) of the definition of cubic splines, we obtain the equation to be satisfied by the G_j's, namely,

$$\begin{aligned} & h_{j-1}G_{j-1} + 2(h_{j-1} + h_j)G_j + h_j G_{j+1} \\ & \qquad = 6\left(\frac{y_{j+1} - y_j}{h_j} - \frac{y_j - y_{j-1}}{h_{j-1}}\right), \end{aligned} \tag{5.8}$$

which is to hold for $j = 2, \ldots, n-2$.

This gives $n-2$ equations in the n unknowns $G_1, \ldots, G_n$, which, with the two additional conditions on G_1 and G_n, gives us an $n \times n$ system of equations. This system is somewhat special, however. It is an example of a tridiagonal system of equations, that is the only nonzero entries are on the main diagonal and the two diagonals immediately above and below it.

For ease of display let us define

$$D_{j-1} = 6\left(\frac{y_{j+1} - y_j}{h_j} - \frac{y_j - y_{j-1}}{h_{j-1}}\right).$$

Then the original $n \times n - 2$ system of equations may be written

$$\begin{bmatrix} h_1 & 2(h_1 + h_2) & h_2 & & & 0 \\ & & & \ddots & & \\ 0 & & & h_{n-2} & 2(h_{n-2} + h_{n-1}) & h_{n-1} \end{bmatrix} \begin{bmatrix} G_1 \\ G_2 \\ \vdots \\ G_{n-1} \\ G_n \end{bmatrix} = \begin{bmatrix} D_1 \\ \vdots \\ D_{n-2} \end{bmatrix}$$

(where the entries that are not filled in are understood to be zeros).

If we impose the additional conditions (iv), (v) or (vi), then the matrix of coefficients becomes

Condition (iv):

$$\begin{bmatrix} 2(h_1 + h_2) & h_2 & & 0 \\ h_2 & 2(h_2 + h_3) & h_3 & \\ & & \ddots & \\ 0 & & h_{n-2} & 2(h_{n-2} + h_{n-1}) \end{bmatrix}$$

Condition (v):

$$\begin{bmatrix} (3h_1 + 2h_2) & h_2 & & 0 \\ h_2 & 2(h_2 + h_3) & h_3 & \\ & & \ddots & \\ 0 & & h_{n-2} & (2h_{n-2} + 3h_{n-1}) \end{bmatrix}$$

Condition (vi):

$$\begin{bmatrix} \frac{(h_1+h_2)(h_1+2h_2)}{h_2} & \frac{h_2^2-h_1^2}{h_2} & 0 & \\ h_2 & 2(h_1 - h_2) & h_3 & \\ & & \ddots & \\ 0 & & \frac{h_{n-2}^2-h_{n-1}^2}{h_{n-2}} & \frac{(h_{n-1}+h_{n-2})(h_{n-1}+2h_{n-2})}{h_{n-2}} \end{bmatrix}$$

It should be noted that since each h_j, $j = 1, \ldots, n$ is positive, all of the systems of equations we must solve are diagonally dominant and hence guaranteed of having a unique solution.

Once we calculate the G_j we then can calculate the coefficients of the $S_j(x)$ using the formulas:

$$\begin{aligned} a_j &= \frac{G_{j+1} - G_j}{6h_j}, \\ b_j &= \frac{G_j}{2}, \\ c_j &= \frac{y_{j+1} - y_j}{h_j} - h_j \frac{2G_j + G_{j+1}}{6}, \qquad \text{and} \\ d_j &= y_j . \end{aligned} \tag{5.9}$$

As an example we shall calculate the cubic spline interpolant for the example above, namely the function

$$f(x) = x^3 + 2x^2 - x + 1 .$$

We remind the reader that the data points chosen were $(-2, 3)$, $(-1, 3)$, $(0, 1)$, $(1, 3)$, and $(2, 15)$, which gives us the values -12, 24 and 60 for D_1, D_2 and D_3, respectively.

For the case of the natural cubic spline, a spline satisfying condition (i), we must solve the system

$$\begin{bmatrix} 4 & 1 & 0 \\ 1 & 4 & 1 \\ 0 & 1 & 4 \end{bmatrix} \begin{bmatrix} G_2 \\ G_3 \\ G_4 \end{bmatrix} = \begin{bmatrix} -12 \\ 24 \\ 60 \end{bmatrix},$$

which gives us the solution

$$G_1 = 0, \quad G_2 = \frac{-27}{7}, \quad G_3 = \frac{24}{7}, \quad G_4 = \frac{99}{7}, \quad G_5 = 0 .$$

To calculate the spline satsifying condition (ii) we must solve the system

$$\begin{bmatrix} 5 & 1 & 0 \\ 1 & 4 & 1 \\ 0 & 1 & 5 \end{bmatrix} \begin{bmatrix} G_2 \\ G_3 \\ G_4 \end{bmatrix} = \begin{bmatrix} -12 \\ 24 \\ 60 \end{bmatrix},$$

which gives us the solution

$$G_1 = \frac{-16}{5}, \quad G_2 = \frac{-16}{5}, \quad G_3 = 4, \quad G_4 = \frac{56}{5}, \quad G_5 = \frac{56}{5} .$$

To calculate the spline satsifying condition (iii) we must solve the system

$$\begin{bmatrix} 6 & 0 & 0 \\ 1 & 4 & 1 \\ 0 & 0 & 6 \end{bmatrix} \begin{bmatrix} G_2 \\ G_3 \\ G_4 \end{bmatrix} = \begin{bmatrix} -12 \\ 24 \\ 60 \end{bmatrix},$$

which gives us the solution

$$G_1 = -8, \quad G_2 = -2, \quad G_3 = 4, \quad , G_4 = 10, \quad G_6 = 16\,.$$

Problem 5.7: Carefully obtain the results of the last paragraph.

We conclude this example by using each of our spline interpolants to estimate $f(0.5)$. Since (0.5) is in the interval $[0, 1]$ we must calculate the cubic polynomial $S_3(x)$. We do this by using the values obtained above for the G's and the formulas in (5.9).

For the natural spline we obtain

$$S_3(x) = \frac{25}{14}\,x^3 + \frac{12}{7}\,x^2 - \frac{3}{2}\,x + 1\,,$$

which gives us the value $S_3(0.5) = 0.901785714\,.$

For the spline satisfying condition (v) we obtain

$$S_3(x) = \frac{6}{5}\,x^3 + 2\,x^2 + -\frac{6}{5}\,x + 1\,,$$

which gives us the value $S_3(0.5) = 1.05\,.$

Finally, for the spline satisfying condition (vi) we obtain

$$S_3(x) = x^3 + 2x^2 - x + 1\,,$$

which gives us the value $S_3(0.5) = 1.125$, which shouldn't be too surprising. As we remarked above, condition (vi) guarantees that if the data originates from a cubic polynomial, then the spline interpolant is that cubic polynomial or one of the Taylor expansions of the cubic polynomial centered about an endpoint of a subinterval.

Problem 5.8: Carefully obtain the results of the last paragraph.

In Exercise 5.12 we ask the reader to write subroutines to do each of the three parts of the cubic spline interpolation problem.

As we see, the spline satisfying condition (vi) gives us the most accurate answers. Indeed, this is usually the case. Because of this, we give the following error estimate.

Theorem 5.4: *Let f be a function which is defined and has a continuous fourth derivative on the interval $[x_1, x_n]$ and is such that*

$$\max_{x_1 \le x \le x_n} |f^{(4)}(x)| \le M$$

where M is some positive constant. If S is the unique cubic spline interplant to f with respect to the points $(x_1, y_1), \ldots, (x_n, y_n)$, which satisfies condition (iii)*, then*

$$\max_{x_1 \le x \le x_n} |f(x) - S(x)| \le \frac{5M}{384} \max_{1 \le j \le n-1} h_j^4 .$$

5.3 Linear Least Square Curve Fitting

The main purpose of this section is to construct the line $y = mx + b$ which best fits a set of data points (x_0, y_0), (x_1, y_1), $\ldots$, (x_n, y_n) in the sense that the error

$$E(m, b) = \sum_{j=0}^{n} (mx_j + b - y_j)^2 \tag{5.10}$$

is a minimum. The solution is called the *least squares solution.* We have pictured this solution where $e_j = |mx_j + b - y_j|$ for $j = 0, 1, 2, 3$. Note that for given m and b, the error is a real number.

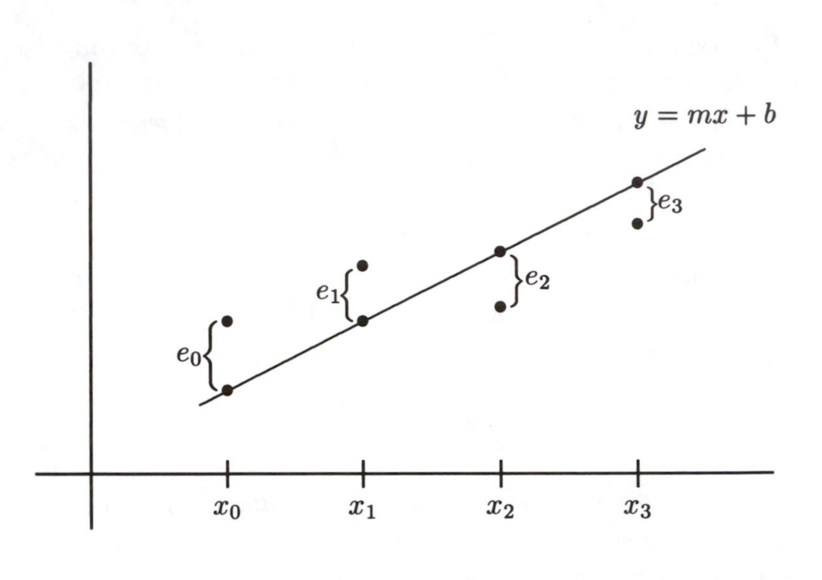

Figure 5.9

Thus, our mathematical problem is that given the data points $(x_0, y_0), (x_1, y_1), \ldots, (x_n, y_n)$ we want to find the values of the parameters m and b which make (5.10) a minimum. A secondary purpose of this section is to illustrate the ideas concerned with least squares problems (See Figure 5.9 above).

The reader might properly ask why we do not minimize the expression

$$F(m, b) = \sum_{j=0}^{n} e_j \,. \tag{5.11}$$

The answer is that (mathematically) this is a very difficult problem while, as we will soon see, it is relatively easy to minimize (5.10). We also note that there is nothing sacred about using a straight line $y = mx + b$. If we suspected that our data came from points involving exponetial growth, we might assume a model $y(x) = Ce^{kx}$ where we seek the values of C and k so as to find the best least squares exponential growth by minimizing

$$G(C,k) = \sum_{j=0}^{n} (Ce^{kx_j} - y_j)^2 .$$

(For a better way to do this problem see Exercise 5.35 below.)

The message or moral is to assume a model that the user believes the data satisfies. As in many problems, it is not immediately obvious that a minimizing solution to our problem exists. We will see that a minimizing solution exists and we will construct the solution.

To find the values of m and b in (5.10) we use the necessary conditions that $\frac{\partial E}{\partial m} = \frac{\partial E}{\partial b} = 0$ at the minimizing solution. Thus,

$$\begin{aligned} 0 &= \frac{\partial E}{\partial m} = \sum_{j=0}^{n} 2x_j(mx_j + b - y_j) \quad \text{and} \\ 0 &= \frac{\partial E}{\partial b} = \sum_{j=0}^{n} 2(mx_j + b - y_j) . \end{aligned} \tag{5.12}$$

The reader should realize that these complicated looking expressions are (only) two linear equations in the two unknowns m and b, that is,

$$\begin{aligned} m\sum x_j^2 + b\sum x_j &= \sum x_j y_j \quad \text{and} \\ m\sum x_j + b(n+1) &= \sum y_j . \end{aligned} \tag{5.13}$$

The determinant

$$D = (n+1)\sum x_j^2 - \left(\sum x_j\right)^2 \neq 0 \tag{5.14}$$

since the x_j's are distinct. Hence (5.12) has a unique solution.

While it it reasonably clear from the geometry of the problem that the values of m and b obtained from (5.13) give the minimizing line we wish to show that this is so from a mathematical point of view, since not all problems yield themselves to geometrical considerations. The results below generalize the result that if $y = f(x)$, $f'(a) = 0$ and $f''(a) > 0$ are sufficient conditions for a local minimum.

We recall the following theorem from calculus

Theorem 5.5: *Let $f(x, y)$ be a function that has continuous partial derivatives f_x, f_y, f_{xx}, f_{yy} and $f_{xy} = f_{yx}$ at and near a point (a, b). We suppose that $f_x(a, b) = 0 = f_y(a, b)$. Then*

a) *if $f_{xx}(a, b) > 0$, $f_{yy}(a, b) > 0$ and $f_{xy}^2(a, b) - f_{xx}(a, b)f_{yy}(a, b) < 0$ then f has a local minimum at (a, b),*

b) *if $f_{xx}(a, b) < 0$, $f_{yy}(a, b) < 0$ and $f_{xy}^2(a, b) - f_{xx}(a, b)f_{yy}(a, b) < 0$, then f has a local maximum at (a, b), and*

c) *if $f_{xy}^2(a, b) - f_{xx}(a, b)f_{yy}(a, b) > 0$, then f has a saddle point at (a, b).*

There are similar results for functions of three or more variables, but they are a little lengthier to write down. The major theoretical ideas leading to Theorem 5.5 are important and extend the optimization results for several independent variables. Since many of these ideas may be important to the readers we now include some of them.

The results of Theorem 5.5 follow from the Taylor series expansion of a real valued function and inequalities involving eigenvalues of symmetric matrices. We assume $X = [x_1, \ldots, x_p]^T$ in $\mathbb{R}^p$ and $F : D \subset \mathbb{R}^p \to \mathbb{R}$ has (at least) continuous second order derivatives. Formally, if X_0 in D we have the Taylor series expansion

$$\begin{aligned} F(X) = F(X_0) &+ [X - X_0]^T F_X \\ &+ \frac{1}{2}[X - X_0]^T F_{XX}[X - X_0] + O(||X - X_0||^3) \end{aligned}$$

where

$$F_X = \left[\frac{\partial F}{\partial X_1}, \ldots, \frac{\partial F}{\partial X_p}\right]^T \quad \text{and} \quad F_{XX} = \left[\frac{\partial^2 F}{\partial X_i \partial X_j}\right].$$

F_X is often called the *gradient* and F_{XX} is the *Hessian.*

As in the case when $p = 1$ the Taylor series expansion immediately leads to the folllowing results

Theorem 5.6: *Under the above assumptions if F has a local (relative) minimum at X_0 then $F_X(X_0) = 0$ and $F_{XX}(X_0) \geq 0$. Conversely, if $F_X(X_0) = 0$ and $F_{XX}(X_0) > 0$ then F has a strict, local minimum at X_0.*

The symbolism $A \geq 0$ for symmetric matrices means that $Y^T AY \geq 0$ for any Y in $\mathbb{R}^p$. In this case we say that A is *nonnegative definite.* It is immediate from Rayleigh-quotient ideas or the Courant-Fisher Theorem that in this case all eigenvalues of A are nonnegative. If $Y^T AY > 0$ for all $Y \neq 0$ in $\mathbb{R}^p$ then A is *positive definite* and all eigenvalues are positive. In the theorem we may replace "maximum" with "minimum" in which case we have $F_{XX}(X_0) \leq 0$ or $F_{XX}(X_0) < 0$. Similarly, $A \leq 0$ indicates that A is *nonpositive definite* (the eigenvalues are nonpositive) and $A < 0$ indicates that A is negative definite (the eigenvalues are negative. Part (a) of Theorem 5.5 is a sufficient condition that $F_{XX}(a, b)$ is positive definite while part (b) of Theorem 5.5 is a sufficient condition that F_{XX} is negative definite.

Problem 5.9: Let $X = [x_1, x_2]^T$,

$$F(X) = x_1^2 + x_2^2 = \begin{bmatrix} x_1 \\ x_2 \end{bmatrix}^T \begin{bmatrix} 1 & 0 \\ 0 & 1 \end{bmatrix} \begin{bmatrix} x_1 \\ x_2 \end{bmatrix}.$$

Verify Theorems 5.5 and 5.6 and the eigenvalue results in the last paragraph. Repeat this verification for the functions $G(X) = -x_1^2 - x_2^2$ and $H(X) = x_1^2 - x_2^2$.

We now return to our problem of finding the minimum of $E(m, 6)$. In this case the vector X in Theorem 5.6 is $[m, b]^T$ and

$F(X)$ is $E(m,b)$. Hence

$$E(m,b) = \sum_{j=0}^{n} (mx_j + b - y_j)^2\,,$$

$$\frac{\partial E}{\partial m} = \sum_{j=0}^{n} 2x_j(mx_j + b - y_j) \quad \text{and}$$

$$\frac{\partial E}{\partial b} = \sum_{j=0}^{n} 2(mx_j + b - y_j)\,,$$

whose critical values are determined from the system (5.13). Thus

$$\frac{\partial^2 E}{\partial m^2} = \sum_{j=0}^{n} 2x_j^2\,, \quad \frac{\partial^2 E}{\partial m \partial b} = \sum_{j=0}^{n} 2x_j \quad \text{and} \quad \frac{\partial^2 E}{\partial b^2} = \sum_{j=0}^{n} 2\,.$$

Since

$$\frac{\partial^2 E}{\partial m^2} = 2\sum_{j=0}^{n} x_j^2 > 0 \quad \text{and} \quad \frac{\partial^2 E}{\partial b^2} = 2(n+1) > 0$$

we have

$$\left(\frac{\partial^2 E}{\partial m \partial b}\right)^2 - \frac{\partial^2 E}{\partial m^2} \cdot \frac{\partial^2 E}{\partial b^2} = 4\left(\sum_{j=0}^{n} x_j\right)^2 - 4(n+1)\sum_{j=0}^{n} x_j^2 < 0$$

by the Cauchy-Schwartz inequality (see Exercises 5.20 and 5.21 below). Thus the unique solution to the system (5.13) does indeed give the minimum value of $E(m,b)$.

For models with many parameters, the above method works in theory. However, the coefficient matrix of the problem often becomes ill-conditioned so that this method is not always practical.

To illustrate the method of least squares we shall find the line $y = mx + b$ that best fits the points $(1,1)$, $(2,2)$, $(3,4)$, $(4,5)$, and

(5, 4). Thus, we consider the error function

$$\begin{aligned} E(m,b) &= \sum_{j=0}^{4} (mx_j + b - y_i)^2 \\ &= (m+b-1)^2 + (2m+b-2)^2 + (3m+b-4)^2 \\ &\quad + (4m+b-5)^2 + (5m+b-4)^2 \,. \end{aligned}$$

Then

$$\begin{aligned} \frac{\partial E}{\partial m} &= 2(m+b-1) + 4(2m+b-2) + 6(3m+b-4) \\ &\quad + 8(4m+b-5) + 10(5m+b-4) \\ &= 110m + 30b - 114 \end{aligned}$$

and

$$\begin{aligned} \frac{\partial E}{\partial b} &= 2(m+b-1) + 2(2m+b-2) + 2(3m+b-4) \\ &\quad + 2(4m+b-5) + 2(5m+b-4) \\ &= 30m + 10b - 32 \,. \end{aligned}$$

Hence, we must see the system of equations

$$\begin{aligned} 110m + 30b &= 114 \\ 30m + 10b &= 32 \,. \end{aligned}$$

Since the determinant of the matrix of coefficients

$$\begin{vmatrix} 110 & 30 \\ 30 & 10 \end{vmatrix} = 200 \neq 0 \,,$$

we obtain $m = 0.9$ and $b = 0.5$. Thus the line of best fit is $y = 0.9x + 0.5$.

Problem 5.10: Repeat the example of the last paragraph with the set of points (1,1), (2,2), (3,4), (4,5), (5,4) and (6,5.9). Can you guess the solution?

5.4 Newton Interpolation Polynomials

In this section we will derive another form of the polynomial that interpolates the $n+1$ given points (x_0, y_0), (x_1, y_1), ..., (x_n, y_n). As we showed in Section 5.1, this polynomial is unique. The virtue of the present approach is that if we really must use polynomials to approximate with, then the Newton polynomials are easier to evaluate than the Lagrange polynomials.

Let y_0, y_1, ..., y_n be a set of values. We define the forward difference operator by

$$\Delta y_k = y_{k+1} - y_k \qquad k = 0, 1, \ldots, n-1\,.$$

We define the second differences by

$$\Delta^2 y_k = \Delta y_{k+1} - \Delta y_k$$

and, in general, for $m \geq 2$ we define

$$\Delta^m y_k = \Delta^{m-1} y_{k+1} - \Delta^{m-1} y_k \qquad k = 0, 1, \ldots, n-1\,. \tag{5.15}$$

To have (5.15) hold for $m = 1$ as well we define

$$\Delta^0 y_k = y_k\,.$$

It is not hard to show, by mathematical induction, that for $m \geq 0$ we have

$$\Delta^m y_k = \sum_{j=0}^{m} (-1)^{m-1} \binom{m}{j} y_{j+k}\,, \tag{5.16}$$

where

$$\binom{m}{n} = \frac{m!}{j!(m-j)!} = \frac{m(m-1)\cdots(m-j+1)}{j!}$$

is the ordinary binomial coefficient (See Exercise 5.24). The proof, in fact, mainly uses one of the major properties of the binomial coefficients, namely $\binom{m+1}{n} = \binom{m}{n} + \binom{m}{n-1}$.) Indeed, for $m = 0$, 1, 2 and 3, we have

$$\begin{aligned}
\Delta^0 y_k &= y_k\,, \\
\Delta^1 y_k &= y_{k+1} - y_k\,, \\
\Delta^2 y_k &= \Delta y_{k+1} - \Delta y_k \\
&= (y_{k+2} - y_{k+1}) - (y_{k+1} - y_k) \\
&= y_{k+2} - 2y_{k+1} + y_k
\end{aligned}$$

and

$$\begin{aligned}
\Delta^3 y_k &= \Delta^2 y_{k+1} - \Delta^2 y_k \\
&= (y_{k+3} - 2y_{k+2} + y_{k+1}) - (y_{k+2} - 2y_{k+1} + y_k) \\
&= y_{k+3} - 3y_{k+2} + 3y_{k+1} - y_k\,.
\end{aligned}$$

One can write these differences in table form. For example, if $y_1 = 5.387$, $y_2 = 7.689$, $y_3 = -3.111$ and $y_4 = 2.337$ we have the following difference table

y	Δy_1	$\Delta^2 y_1$	$\Delta^3 y_1$
5.387			
	2.303		
7.689		−13.102	
	−10.8		29.35.
−3.111		16.248	
	5.448		
2.337			

These difference tables can be useful in trying to determine the functional form the data might have arisen from as well as in the determination of error propogation in tables. Those ideas are outside of our present purpose, however, and so we shall not consider them.

We now discuss the idea of divided differences. Suppose we have the data points (x_0, y_0), ..., (x_n, y_n). We define the first order divided difference by

$$\frac{y_{k+1} - y_k}{x_{k+1} - x_k} = \frac{\Delta y_k}{\Delta x_k}$$

and denote it by $\Delta[x_{k+1}x_k]$. Note that since both Δy_k and Δx_k are antisymmetric we have

$$\Delta[x_{k+1}x_k] = \Delta[x_k x_{k+1}] .$$

If we define

$$\Delta^0[x_k] = y_k ,$$

then for $m \geq 0$ we define the mth order divided difference to be

$$\Delta^m[x_{k+m}x_k] = \frac{\Delta^{m-1}[x_{m+k}x_{k+1}] - \Delta^{m-1}[x_{m+k-1}x_k]}{x_{m+k} - x_k} . \tag{5.17}$$

For example,

$$\Delta^2[x_{k+2}x_k] = \frac{\Delta[x_{k+2}x_{k+1}] - \Delta[x_{k+1}x_k]}{x_{k+2} - x_k} .$$

Just as the forward differences for a set of values $\{y_0, \ldots, y_n\}$ can be written solely in terms of the appropriate y-values no matter what the order desired, we can write the divided differences of any order solely in terms of the appropriate x's and y's. For example, we have, for order $m = 1$ and $m = 2$,

$$\begin{aligned} \Delta[x_{k+1}x_k] &= \frac{y_{k+1} - y_k}{x_{k+1} - x_k} \\ &= \frac{y_{k+1}}{x_{k+1} - x_k} + \frac{y_k}{x_k - x_{k+1}} \end{aligned}$$

and

$$\begin{aligned}
\Delta^2[x_{k+2}x_k] &= \frac{\Delta[x_{k+2}x_k] - \Delta[x_{k+1}x_k]}{x_{k+2} - x_k} \\
&= \frac{1}{x_{k+2} - x_k}\left(\frac{y_{k+2} - y_{k+1}}{x_{k+2} - x_{k+1}} - \frac{y_{k+1} - y_k}{x_{k+1} - x_k}\right) \\
&= \frac{y_{k+2}}{(x_{k+2} - x_k)(x_{k+2} - x_{k+1})} \\
&\quad + \frac{y_{k+1}}{(x_{k+1} - x_k)(x_{k+1} - x_{k+2})} \\
&\quad + \frac{y_k}{(x_k - x_{k+1})(x_k - x_{k+2})}\,.
\end{aligned}$$

In general, an easy induction argument shows that, for $m \geq 1$, we have

$$\Delta^m[x_{k+m}x_k] = \sum_{j=0}^{m} y_{j+k} \prod_{\substack{i=0 \\ i\neq j}} (x_{k+j} - x_{k+1})^{-1}\,. \tag{5.18}$$

(See Exercise 5.28a.)

We now are in a position to produce the Newton interpolation polynomial. Here we are looking for a polynomial of the form

$$\begin{aligned}
p(x) = c_0 &+ c_1(x - x_0) + c_2(x - x_0)(x - x_1) \\
&+ c_3(x - x_0)(x - x_1)(x - x_2) + \cdots \\
&+ c_n(x - x_0)(x - x_1)(x - x_2)\cdots(x - x_{n-1})\,.
\end{aligned}$$

The value of this form of the interpolating polynomial comes from its computational ease. Indeed

$$p(x) = c_0 + (x - x_0)[(c_1 + (x - x_1)(c_2 + \cdots + c_n(x - x_{n-1})\cdots)]$$

so that the computational algorithm would look like

$$p = c_j + (x - x_j)p$$

where the intial value of p is c_n. The problem, then, is to compute the coefficients c_j, $j = 0, 1, \ldots, n$.

To do this note that if we are to have $p(x_j) = y_k$, for $j = 0, 1, \ldots, n$, then $y_0 = p(x_0)$ implies that

$$y_0 = c_0 .$$

If we let $x = x_1$, then we have

$$y_1 = y_0 + c_1(x_1 - x_0)$$

or

$$c_1 = \frac{y_1 - y_0}{x_1 - x_0} = \Delta[x_1 x_0] .$$

If we let $x = x_2$, we get

$$y_2 = y_0 + \Delta[x_1 x_0](x_2 - x_0) + c_2(x_2 - x_0)(x_2 - x_1)$$

or

$$\begin{aligned} c_2 &= \frac{y_2 - y_0 - \Delta[x_1 x_0](x_2 - x_0)}{(x_2 - x_0)(x_2 - x_1)} \\ &= \frac{y_2 - y_1 + \Delta[x_1 x_0](x_1 - x_0) - \Delta[x_1 x_0](x_2 - x_0)}{(x_2 - x_0)(x_2 - x_1)} \\ &= \frac{(y_2 - y_1) + \Delta[x_1 x_0](x_1 - x_2)}{(x_2 - x_0)(x_2 - x_1)} \\ &= \frac{\Delta[x_2 x_1] - \Delta[x_1 x_0]}{x_2 - x_0} \\ &= \Delta^2[x_2 x_0] , \end{aligned}$$

since $y_0 = y_1 - \Delta[x_1 x_0](x_1 - x_0)$. Likewise one may show that for $m = 0, 1, \ldots, n$, we have

$$c_m = \Delta^m[x_m x_0] .$$

Thus, we have the Newton Interpolation Polynomial

$$p(x) = \Delta^0[x_0] + \Delta[x_1x_0](x - x_0) + \cdots \\ + \Delta^n[x_nx_0](x - x_0)\cdots(x - x_{n-1})\,. \tag{5.19}$$

As an example, consider the data points $(-1,-3)$, $(4,2)$, $(3,-1)$, $(1,-1)$. To compute the divided differences we use a divided difference table

x	y	$\Delta[x_{k+1}x_k]$	$\Delta^2[x_{k+2}x_k]$	$\Delta^3[x_{k+3}x_k]$
-1	-3			
		1		
4	2		$\frac{1}{2}$	
		3		$\frac{1}{4}$
3	-1		1	
		0		
1	-1			

Thus the Newton polynomial is, by (5.19),

$$p(x) = -3 + 1(x+1) + 0.5(x+1)(x-4) \\ + 0.25(x+1)(x-4)(x-3)\,.$$

Since $p(x) = -3 + (x+1)\big(1 + (x-4)\big(0.5 + 0.25(x-3)\big)\big)$ we may interpolate at $x = 2$ relatively easily to get

$$p(2) = -3 + 3\big(1 + (-1)(0.5 + 0.25(0.1))\big) = 0.75\,.$$

It should be noted that the calculations involving the Lagrange interpolation polynomial are not this easy.

Finally, we note what happens when the x values are equally spaced. Suppose $x_{k+1} - x_k = h$, $k = 0, 1, \ldots, n-1$. Then $x_k =$

$x_0 + kh$, $k = 0, 1, \ldots, n$, and by (5.18),

$$\begin{aligned}
&\Delta^m[x_{k+m}x_k] \\
&\qquad = \sum_{j=0}^{m} y_{j+k} \prod_{\substack{i=0 \\ i \neq j}}^{m} (x_{k+j} - x_{k+1})^{-1} \\
&\qquad = \sum_{j=0}^{m} y_{j+k} \prod_{\substack{i=0 \\ i \neq j}}^{m} \left[x_0 + (k+j)h - x_0 - (k+i)h\right]^{-1} \\
&\qquad = \sum_{j=0}^{m} y_{j+k} \prod_{\substack{i=0 \\ i \neq j}}^{m} (j-i)^{-1} h \\
&\qquad = \frac{h^{-m}}{m!} \sum_{j=0}^{m} (-1)^{m-j} \binom{m}{j} y_{j+k} \\
&\qquad = \frac{\Delta^m y_k}{h^m m!},
\end{aligned} \tag{5.20}$$

by (5.16). We have

$$\begin{aligned}
&(x - x_0)(x - x_1) \cdots (x - x_{k-1}) \\
&\qquad = (x - x_0)(x - x_0 - h) \cdots (x - x_0 - (k-1)h) \\
&\qquad = h \frac{x - x_0}{h} h \left(\frac{x - x_0}{h} - 1 \right) \cdots h \left(\frac{x - x_0}{h} - k + 1 \right) \\
&\qquad = h^k k! \binom{\frac{x - x_0}{h}}{k},
\end{aligned} \tag{5.21}$$

where

$$\binom{z}{k} = \frac{z(z-1)\cdots(z-k+1)}{k!}.$$

Then by (5.19), (5.20) and (5.21), we have

$$\begin{aligned}
p(x) &= y_0 + \sum_{j=1}^{n} \Delta^j [x_j x_0] \prod_{i=0}^{j-1} (x - x_i) \\
&= y_0 + \sum_{j=1}^{n} \frac{\Delta^j y_0}{h^j j!} \cdot h^j j! \binom{\frac{x-x_0}{h}}{j} \\
&= y_0 + \sum_{j=1}^{n} \Delta^j y_0 \binom{\frac{x-x_0}{h}}{j},
\end{aligned}$$

which is called the Newton Binomial Interpolation Polynomial centered about $\dfrac{x - x_0}{h}$.

Exercise Set 5

1. Graph $f(x) = x^2 - x$ on $[-2, 2]$. Choose three points $(x_0, f(x_0))$, $(x_1, f(x_1))$, and $(x_2, f(x_2))$ so that $\frac{1}{2} < x_0 < x_1 < x_2 < 1$. What is the interpolating polynomial of degree two for these points? Are you satisfied with the interpolating polynomial?

2. Draw a smooth curve containing the points $(-2, f(-2))$, $(-\frac{3}{2}, f(-\frac{3}{2}))$, $\frac{3}{2}, f(\frac{3}{2}))$, and $(2, f(2))$ where $f(x) = x^2 - x$. What is the interpolating polynomial of these points? What is the moral?

3. Repeat the proof of Theorem 5.2 for $n = 2$ and $S = \{(0, 5), (1, 3), (2, 3)\}$. That is, show that there is a unique polynomial of degree less than or equal to two, which interpolates these points by constructing the Vandermonde determinant.

4. Find the interpolating polynomial for the set S in Exercise 3 using Lagrange interpolation.

5. Let $S = \{(0,5),(1,3),(3,2),(3,5)\}$. Find the interpolating polynomial for S. Why is your answer the same as in Exercise 4?

6. If $\ell_j(x)$ is as given in (5.2) find $f(x) = \sum_{j=0}^{n} \ell_j(x)$. The hint is to set $y_j = 1$, $j = 0,1,\ldots,n$ in (5.3).

7. Prove (5.4). As a hint we note that if $x = x_i$, $i = 0,1,\ldots,n$ then (5.4) holds immediately. For x in $[a,b]$, $x \neq x_i$ for $i = 0,1,\ldots,n$ define

$$E(x) = f(x) - p(x) = (x - x_0)(x - x_1)\cdots(x - x_n)g(x)$$

and

$$w(t) = f(t) - p(t) = (t - x_0)(t - x_1)\cdots(t - x_n)g(x)$$

where $p(t)$ is the interpolating polynomial and $g(x)$ is a function to be determined so that $E(x) = 0$. We note that $w(t) = 0$ for the $n+2$ points, $t = x_0, x_1, \ldots, x_n$ and $t = x$. Furthermore, $w^{(n+1)}(t)$ is continuous in $[a,b]$. By Rolle's Theorem, $w'(t)$ vanishes $n+1$ times on $(a,b]$, $w''(t)$ vanishes n times on (a,b), ..., $w^{n+1}(t)$ vanishes once on (a,b). Thus, for ξ in (a,b)

$$w^{(n+1)}(\xi) = f^{(n+1)}(\xi) - (n+1)!g(x) = 0$$

since $p(t)$ is a polynomial of degree at most n and

$$(t - x_0)(t - x_1)\cdots(t - x_n) = t^{n+1} - p_1(t)$$

where $p_1(t)$ is also a polynomial of at most n.

8. Find the error using (5.4) for Exercise 3. Are you surprised by this result?

9. Let P be the linear function interpolating f at a and b. Show that if $|f''(x)| \leq M$ on $[a, b]$ then

$$|f(x) - P(x)| \leq \frac{1}{8}M(b-a)^2S$$

for all x in $[a, b]$.

10. Consider the function $f(x) = \dfrac{x}{1+x^2}$.

    a) Find $p_1(x)$, the polynomial of degree $\leq$ 4 that passes through the points $(-1, f(-1))$, $(-\frac{1}{2}, f(-\frac{1}{2}))$, $(0, f(0))$, $(\frac{1}{2}, f(\frac{1}{2}))$ and $(1, f(1))$. Calculate $f(\frac{1}{3}) - p_1(\frac{1}{3})$ and $f(3) - p_1(3)$.

    b) Let $x_k = \cos\frac{(2k-1)\pi}{10}$ for $k = 1, 2, 3, 4$ and 5. (These points are examples of what are called Tchebycheff nodes.) Let $p_2(x)$ be the polynomial of degree ≤ 4 that passes through the points $(x_1, f(x_1))$, $(x_2, f(x_2))$, $(x_3, f(x_3))$, $(x_4, f(x_4))$, $(x_5, f(x_5))$. Calculate $f(\frac{1}{3}) - p_2(\frac{1}{3})$ at $f(3) - p_2(3)$. (HINT: some trigonometric identities may prove useful.)

    c) Graph f, p_1 and p_2 on the same set of axes.

11. We have a table of values of e^x, say $e^0 = 1$, $e^{0.1} = 1.1052$, $e^{0.2} = 1.2214$ and $e^{0.3} = 1.3499$. Suppose we want to interpolate the value of $e^{0.15}$ by the Lagrange formula? What can we expect for the minimum and maximum errors? Use a cubic polynomial to interpolate these values and give a value for $e^{0.15}$.

12. As noted above in the material on cubic splines there are three parts to the spline interpolation problem:

    1) the computation of $G_1, \ldots, G_n$,
    2) the computation of a_j, b_j, c_j and d_j for $j = 1, 2, \ldots, n$, and have the construction of $S_j(x)$ for $j = 1, \ldots, n$, and
    3) the evaluation of $S(x)$, and hence the appropriate $S_j(x)$, at some point in the interval $[x_1, x_n]$.

    Write subroutines that do each of these three parts for each of the three types of cubic splines considered in the text. Note that the consideration of each type of spline only effects part (1).

13. Apply subroutines from Exercise 12 to the following problem.

    Consider the set of points $(0, e^0)$, $(0.5, e^{0.5})$, $(1, e^1)$, $(1.5, e^{1.5})$, $(2, e^2)$, $(2.5, e^{2.5})$, and $(3, e^3)$. Fit a natural cubic spline to this set of data and give the coefficients of the interpolating cubic polynomials. Then calculate, from your spline interpolation, the values when $x = 0.25$, 0.75, 1.25, 1.75, 2.25, and 2.75. Compare these with the true values of $e^{0.25}$, $e^{0.75}$, $e^{1.25}$, $e^{1.75}$, $e^{2.25}$, and $e^{2.75}$. (HINT: you only have to do the subroutines for parts (1) and (2) of Exercise 12 once.)

14. Consider the function $f(x) = \dfrac{1}{1 + 4x^2}$.

    a) Fit a natural cubic spline on the interval $[-1, 1]$ using 4 points $(-1, f(-1))$, $(-\frac{1}{3}, f(-\frac{1}{3}))$, $(\frac{1}{3}, f(\frac{1}{3}))$, $(1, f(1))$.

    b) Let $x_k = \cos \frac{2k-1}{8}\pi$, $k = 1, 2, 3$, and 4. Fit a natural cubic spline on this interval $[-1, 1]$ using the points $(x_1, f(x_1))$, $(x_2, f(x_2))$, $(x_3, f(x_3))$, $(x_4, f(x_4))$. (HINT: some trigonometric identities may prove useful.)

15. Let $(x_1, y_1), \ldots, (x_{n+1}, y_{n+1})$ be $n+1$ points such that $x_1 < x_2 < \cdots < x_n < x_{n+1}$.

    a) Let $f(x)$ be continuously differentiable on $[x_1, x_{n+1}]$ and such that $f(x_j) = y_j$ for $j = 1, \ldots, n+1$. If $S(x)$ is the linear spline that interpolates these $n+1$ points, then

    $$\int_{x_1}^{x_{n+1}} (S'(x))^2 \, dx \leq \int_{x_1}^{x_{n+1}} (f'(x))^2 \, dx .$$

    (HINT: integrate by parts.)

    b) Let $f(x)$ be a function whose second derivative is continous on $[x_1, x_{n+1}]$ and such that $f(x_j) = y_j$ for $j = 1, \ldots, n+1$. If $S(x)$ is the natural cubic spline that interpolates these $n+1$ points, then

    $$\int_{x_1}^{x_{n+1}} (S''(x))^2 \, dx \leq \int_{x_1}^{x_{n+1}} (f''(x))^2 \, dx .$$

    (HINT: let $g = f - S$ so that $g(x_j) = 0$ for $j = 1, \ldots, n+1$. Integrate by parts and use the fact that $S''(x_1) = S''(x_{n+1}) = 0$.)

16. Calculate the natural cubic spline that interpolates the three points (x_1, y_1), (x_2, y_2) and (x_3, y_3).

17. Determine a, b, c and d so that

    $$S(x) = \begin{cases} x^3 + 1 & -1 \leq x \leq 0 \\ ax^3 + bx^2 + cx + d & 0 \leq x \leq 2 \end{cases}$$

    is a natural cubic spline.

18. Draw a free-form curve on graph paper. Select 25 points on this curve and compute the cubic spline that takes on these values. Select 5 other points on your curve and check the values given by the spline interpolant against the actual values on the curve.

19. Let

$$f(x) = \frac{1}{1+25x^2}$$

(which, for historical reasons is called Runge's function).

a) Use the Lagrange interpolation formula to find the $20th$ degree polynomial which goes through the 21 points $(-2, f(-2))$, $(-1.8, f(-1.8))$, $\ldots$, $(1.8, f(1.8))$, $(2, f(2))$.

b) Determine the natural cubic spline that interpolates $f(x)$ through these points given in (a).

c) Compare the accuracy of the two results.

20. Show that (5.14) of Section 5.3 is true. You may use the Cauchy-Schwartz inequality $|(x, y)| \leq ||x||\, ||y||$ with $x = [x_0, x_1, \ldots, x_n]^T$ and $y = [1, 1, \ldots, 1]^T$ with the usual norm and inner product. The Cauchy-Schwartz inequality may be proven by starting with the fact that $0 \leq \sum(x_j + \lambda y_j)^2$ for any λ. This is a quadratic inequality in λ and hence the discriminant is nonnegative. The discriminant is zero if and only if $x_j + \lambda y_j = 0$ for each j and some real λ.

21. Use the data points in the example in Section 5.3 to find a least squares fit for the model $y = a_0 + a_1 x + a_2 x^2$.

22. Show that (5.14) is true by defining

$$\overline{x} = \frac{1}{n+1} \sum_{i=0}^{n} x_i$$

and expanding

$$(n+1) \sum_{i=0}^{n} (x_i - \overline{x})^2 .$$

23. In this problem we shall give a constructive polynomial approximation theorem which is due to S. Bernstein.

    a) If the binomial coefficients are defined, as usual, by

    $$\binom{n}{k} = \frac{n!}{k!(n-k)!},$$

    then show that

    $$\binom{n-1}{k-1} = \frac{k}{n}\binom{n}{k} \quad \text{and} \quad \binom{n-2}{k-2} = \frac{k(k-1)}{n(n-1)}\binom{n}{k} .$$

    b) By recalling the binomial theorem show that for any x we have

    $$1 = \sum_{k=0}^{n} \binom{n}{k} x^k (1-x)^{n-k}.$$

    c) Use the result of part (b) to show that for any x we have

    $$x = \sum_{k=0}^{n} \frac{k}{n} \binom{n}{k} x^k (1-x)^{n-k}.$$

    (HINT: multiply the result in part (b) by x and rearrange the sum.)

d) Similarly show that by using the results of parts (b) and (c) one has

$$\frac{1}{n}x(1-x) = \sum_{k=0}^{n} \left(x - \frac{k}{n}\right)^2 \binom{n}{k} x^k (1-x)^{n-k}.$$

e) If f is a function defined on the interval $[0, 1]$, then the nth Bernstein polynomial for f is defined to be

$$B_n(x, f) = \sum_{k=0}^{n} f(\frac{k}{n}) \binom{n}{k} x^k (1-x)^{n-k}.$$

Show that

$$B_n(x, 1) = 1\,, \qquad B_n(x, x) = x$$

and

$$B_n(x, x^2) = \left(1 - \frac{1}{n}\right) x^2 + \frac{1}{n} x\,.$$

Find $B_3(x, \sin x)$ in increasing powers of x.

f) Show that if $B_n(x, f)$ is the nth Bernstein polynomial for f, then

$$f(x) - B_n(x, f) = \sum_{k=0}^{n} [f(x) - f(\frac{k}{n})] \binom{n}{k} x^k (1-x)^{n-k}.$$

(HINT: use part (b).)

g) Suppose f is continuous on $[0,1]$. Then prove that for any $\epsilon > 0$ there is an n, independent of x, such that

$$|f(x) - B_n(x,f)| < \epsilon .$$

(HINT: this uses the concept of uniform continuity. The value of n needed is defined by

$$n = \sup\left(\delta^{-4}, \frac{4M^2}{\epsilon^2}\right) ,$$

where $|f(x)| \leq M$ for x in $[0,1]$ and $|x - \frac{k}{n}| < \epsilon$ implies $|f(x) - f(\frac{k}{n})| < \frac{\epsilon}{2}$. The result of part (d) will be needed. One breaks the sum from part (f) into two sums: one over those k such that $|x - \frac{k}{n}| \leq n^{-1/4}$ and the other over the other k.)

h) Find the Bernstein polynomial $B_n(x)$ that approximates e^x to within 0.1 on the interval $[0,1]$.

i) Find the Bernstein polynomial $B_n(x)$ that approximates $|2x-1|$ to within 0.1 on the interval $[0,1]$.

24. Prove equation (5.16) of Section 5.4.

25. For the forward difference operator show that if $\{x_1, \ldots, x_n\}$ and $\{y_1, \ldots, y_n\}$ are two sets of data, then

a) $\Delta^m(x_k + y_k) = \Delta^m x_k + \Delta^m y_k$,

b) $\Delta(x_k y_k) = x_k \Delta y_k + y_{k+1} \Delta x_k$,

and

c) $\Delta(x_k y_k) = x_k \Delta y_k + y_k \Delta x_k + \Delta x_k \Delta y_k$.

26. This exercise shows how the forward difference operator behaves under error propogation. Denote by y the correct data values and by y' the incorrect data values. Let

$$e_k = y'_k - y_k\,, \quad k = 1, 2, \ldots, n+1\,.$$

a) Show that for $k = 1, \ldots, n$ we have

$$\Delta^k y'_1 = \Delta^k y_1 + \Delta^k e_1\,.$$

b) Let $e = \max_{i \le k \le n} |e_k|$. Show that $|\Delta e_k| \le 2e$.

c) Show that for $k = 1, \ldots, n$ we have $|\Delta^k e_1| \le 2^k e$.

d) Show that for $k = 1, \ldots, n$ we have

$$|\Delta^k y'_1 - \Delta^k y_1| \le 2^k e\,.$$

27. If $p(x) = a_n x^n + \cdots + a_1 x + a_0$ is a polynomial of degree n, show that

$$\Delta^n p(x) = n! a_n h^n \quad \text{and} \quad \Delta^m p(x) = 0 \text{ for } \; m > n$$

for any value of x, where the x values are spaced with a step size of h.

28. a) Prove equation (5.18) of Section 5.4.

b) Show that the order of the points is immaterial in the computation of $\Delta^m[x_{k+m} x_k]$.

Computational Problems

29. a) Form the divided difference table for the data points

$$\{(3,5),(-1,3),(7,2),(5,3),(6,-1),(-2,3),(4,1)\}.$$

b) Find the Newton interpolation polynomial for the data points of part (a).

30. a) Form the divided difference table for the data points

$$\{(1.5,\sqrt{1.5}),(1.5,\sqrt{1.6}),(1.7,\sqrt{1.7}),(1.8,\sqrt{1.8}),\\ (1.9,\sqrt{1.9}),(2,\sqrt{2})\}.$$

b) Find the Newton interpolation polynomial for the data points in part (a).

c) Compare the values of the polynomial of part (b) at $x =$ 1.55, 1.75 and 2.15 with the corresponding values of $\sqrt{x}$.

31. Given the values $(1, \ln 1)$, $(2, \ln 2)$, $\ldots$, $(10, \ln 10)$, find the Newton binomial interpolation polynomial that fits these data points. Compare the values of this polynomial at $x = 5.75$ with the value of $\ln 5.75$.

32. Find the best fitting lines, in the least squares sense, that fit the following set of data points:

a) $\{(3,1),(2,-1),(6,10),(-1,3),(4,2)\}$

b) $\{(-1,3),(1,1),(0,3),(-2,1),(4,10),(8,-1)\}$.

33. Find the best fitting parabola, that is, a curve of the form $y = ax^2 + bx + c$, in the least squares sense, for the data points in Exercise 5.31.

34. Find the best fitting plane, that is, a curve of the form $z = ax + by + c$, in the least squares sense, that fits the following data points:

$$\{(2,-1,3),(-1,2,3),(6,-1,7),(8,-1,1),(0,0,3)\}.$$

35. To fit an exponential curve $y = Ae^{Bx}$, one can make a change of variables by taking logarithms of both sides to get

$$\log y = Bx + \log A$$

and the letting $z = \log y$. We then fit a line through the data points

$$\{(x_1, \log y_1),(x_2, \log y_2),\ldots,(x_n, \log y_n)\}$$

to find B and $\log A$ and from this we get the exponential curve. Try to fit an exponential curve through the data points:

$$\{(-1,3),(-2,2),(-3,1.5),(1,9.5),(2,15),(3,24)\}.$$

36. Show that if we try to best fit, in the least squares sense, a line through two data points, we get the line that goes through those points.
37. To fit a power curve $y = Ax^B$, one can proceed as in Exercise 5.35 to replace the curve by

$$\log y = B \log x + \log A\,.$$

If we let $u = \log y$ and $v = \log x$, we can then try to fit a line through the data points

$$\{(\log x_1, \log y_1),(\log x_2, \log y_2),\ldots,(\log x_n, \log y_n)\}$$

to find B and $\log A$ and from this we get the power curve. Try to fit a power curve to the data points

$$\{(2,6.9),(3,11.2),(4,15.8),(5,20.7),(6,25.75)\}.$$

6 *Numerical Integration and Differentiation*

The purpose of this chapter is to consider how to numerically solve the two basic problems in calculus which are to find the derivative at a point (rate of change) and to find the definite integral (generalized area) under a curve. The major tools are the interpolating polynomials of the last chapter. Specifically, the *problem* is that given the set S of data points in Chapter 5, we seek an approximation for the derivative at a point x or the definite integral of f on $[a, b]$. Once again, we will see that the estimation of errors is a major part of these topics.

Our first section on numerical differentiation is very brief. The student should realize that numerical differentiation is a chaotic and unsatisfactory task. Small errors in the data points lead to large errors in the derivatives. Our major result is to give numerical formulas of $f'(x_0)$ and $f''(x_0)$ and the associated errors. The remainder of the chapter is devoted to numerical integration. These results are much more satisfactory since integration is a smoothing process.

In Section 6.2, we review the basic theory of the definite integral of a function f on $[a, b]$. While this materical is optional it is important to understand that under week hypotheses quadrature rules lie between the upper and lower Riemann sums which are approaching the same value (the definite integral) as the norm of the partition goes to zero. In Section 6.3 we introduce the basic quadrature rules to approximate the definite integral of f on $[a, b]$ and find the corresponding error bounds. These rules are usually called the rectangular rule, the trapezoid rule and Simpson's rule. In Section 6.4, we introduce a class of quadrature rules known as Newton-Cotes Closed Form rules. Finally, in Section 6.5, we present Romberg integration. This is an application of a Richardson procedure which (almost) allows us to obtain something for nothing in that relatively small amounts of additional effort lead to substantial improvement in our results. The Richardson procedure which the first author has used in his own research will also be discussed in more detail in Chapter 7.

6.1 Numerical Differentiation

Assuming, as in Chapter 5, that S denotes the set of $n+1$ points $(x_0, y_0), (x_1, y_1), \ldots, (x_n, y_n)$ where $x_0 < x_1 < x_n$, the *problem* is to find the value of $f'(x)$ and possibly higher derivatives. The usual and often best solution is to pass a smooth curve, $y = g(x)$, through a subset of S and then define the numerical approximation of f' at x as $g'(x)$. As we have noted, this is an ill-conditioned problem in that small changes in the value of $\{y_k\}$ are often magnified and result in large errors in the calculated values of $f'(x)$. In fact, we have seen in the last chapter that it is often difficult to find good values of the function, let alone the derivative. The reader will recall that one of the main advantages of spline interpolation is that it allows us to find curves whose derivatives appear reasonable.

It is surprising, but true, that polynomial interpolation can give such poor results for the values of the derivative. For example, we recall Figure 5.4, denoted as Figure 6.1 below, where P_1, P_2, P_3, and P_4 are points from the function $y = f(x)$ which have been interpolated by the cubic polynomial $y = P_3(x)$. The reader can easily see that at most points in the interval $[x_3, x_4]$, the value $f'(x)$ is very different from $P_3'(x)$. Similarly, the reader can jiggle the points P_3 and P_4 on the lines $y = x_3$ and $y = x_4$ respectively, to see that the slopes of the chord connecting P_3 and P_4 vary greatly with minor jiggling.

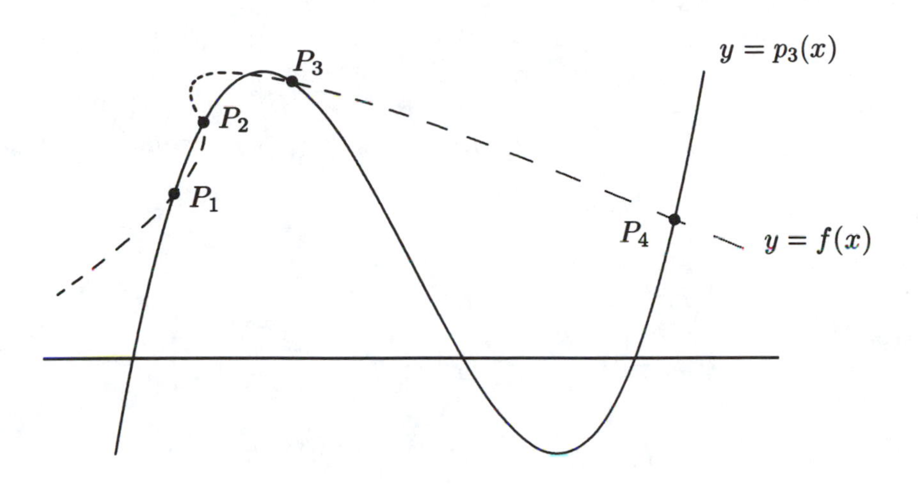

Figure 6.1

Problem 6.1: Make up an example which illustrates the results in the last sentence.

With apologies to the reader we now change our notation somewhat. This will allow us to make our derivations easier and results clearer. We now assume that an interior point x_0 is the point at which we wish to find $f'(x_0)$, $f''(x_0)$, etc. Hence we assume the points

$$(x_{-k}, f_{-k}), (x_{-k+1}, f_{-k+1}), \ldots,$$
$$(x_{-1}, f_{-1}), (x_0, f_0), (x_1, f_1), \ldots, (x_k, f_k)$$

are in S (for some convenient $k > 0$) with $x_{i+1} - x_i = h$ and $f_j = f(x_j)$ for $j = -k, -k+1, \ldots, k$. The *problem* we now wish to address is to give numerical formulas for $f'(x_0)$ and $f''(x_0)$ in terms of the data points of S where (x_0, f_0) is a point in S and the x values of S are equally spaced. Assuming f is suitably differentiable we begin with the formulas

$$f_1 = f(x_1) = f(x_0 + h) \tag{6.1}$$
$$= f_0 + hf_0' + \frac{h^2}{2}f_0'' + \frac{h^3}{6}f_0''' + \frac{h^4}{24}f_0^{(iv)} + \cdots$$

and

$$f_{-1} = f(x_{-1}) \tag{6.2}$$
$$= f(x_0 - h) = f_0 - hf_0' + \frac{h^2}{2}f_0'' - \frac{h^3}{6}f_0''' + \frac{h^4}{24}f_0^{(iv)} + \cdots$$

where $f_0' = f'(x_0)$, $f_0'' = f''(x_0)$, etc. Using (6.1) we write

$$f_0' = \frac{f_1 - f_0}{h} + O(h) \tag{6.3}$$

to indicate that our numerical approximation to $f'(x_0)$ is given by the number $\frac{f_1 - f_0}{h}$ and has an error proportional to h as $h \to 0$. That is, there exists a constant M which is positive and independent of h for h small, such that

$$\left| f'(x_0) - \frac{f(x_1) - f(x_0)}{h} \right| \le Mh\,.$$

It is important to note that if we can find new data points with $x_{i+1} - x_i = \frac{h}{2}$ the error in (6.3) would "be halved" for h small.

We might also subtract (6.2) from (6.1) to obtain

$$f_0' = \frac{f_1 - f_{-1}}{2h} + O(h^2)\,. \tag{6.4}$$

Thus, if we approximate $f'(x_0)$ by what is usually called the *central difference formula*, $\frac{f_1 - f_{-1}}{2h}$, for some $M_1 > 0$ and independent of h, we have

$$\left| f(x_0) - \frac{f(x_1) - f(x_{-1})}{2h} \right| \le M_1 h^2\,.$$

This time, new data points with $\frac{h}{2}$ replacing h gives an error "one quarter" of the original error.

We can also find approximate values of $f''(x_0)$ using this technique. If we add (6.1) to (6.2) we get

$$f_1 + f_{-1} = 2f_0 + h^2 f_0'' + \frac{h^4}{12} f_0^{(iv)} + \cdots$$

or

$$f_0'' = \frac{f_1 - 2f_0 + f_{-1}}{h^2} + O(h^2)\,. \tag{6.5}$$

Note that new data points with $\frac{h}{2}$ replacing h gives an error one quarter the original error. The number $\frac{f_1 - 2f_0 + f_{-1}}{h^2}$ is usually called the *central difference formula* for the second derivative.

Problem 6.2: Assuming $k > 1$ in the above, find formulas for $f'(x_1)$, $f''(x_1)$ similar to (6.4) and (6.5).

Other expressions can be similarly derived. If we replace (6.1) and (6.2) by

$$f_2 = f_0 + 2hf_0' + \frac{(2h)^2}{2} f_0'' + \frac{(2h)^3}{6} f_0''' + \frac{(2h)^4}{24} f_0^{(iv)} + \cdots \tag{6.6}$$

and

$$f_{-2} = f_0 - 2hf_0' + \frac{(2h)^2}{2} f_0'' - \frac{(2h)^3}{6} f_0''' + \frac{(2h)^4}{24} f_0^{(iv)} + \cdots \tag{6.7}$$

we may combine (6.6) and (6.7) with (6.1) and (6.2) to obtain a variety of new formulas (see Exercise 6.1).

6.2 The Definite Integral

We now turn to the problem of numerical quadrature, that is, the problem of finding numerical approximations for definite integrals. Before we begin with the numerical algorithms we must discuss some of the theory of the definite integral.

We begin by recalling the Fundamental Theorem of Calculus: If $F'(x) = f(x)$, then

$$\int_a^b f(x)\,dx = F(b) - F(a)\,.$$

Thus, in theory, we can calculate any definite integral of a function that possesses an *antiderivative.* The problems arise when we run across functions that do not possess antiderivatives that are easily calculated. For example, in this category we would place functions such as $\sin x^2$, e^{x^2}, elliptic functions, Bessel functions and, indeed, many of the functions which arise in applications. If this is the case, that is, we cannot easily calculate the antiderivative, then we must approximate the definite integral. To see how we would go about doing this we must turn to the definition of the definite integral.

To define the definite integral of $f(x)$ over the interval $[a, b]$ we must first define a partition, P, of the interval $[a, b]$. A set of points $\{x_0, x_1, \ldots, x_{n-1}, x_n\}$ is said to be a *partition* of $[a, b]$ if

$$a = x_0 < x_1 < \cdots < x_{n-1} < x_n = b\,.$$

We define a *Riemann sum* of $f(x)$ over $[a, b]$ as the sum

$$S(P; f) = \sum_{i=0}^{n-1} f(x_i^*)(x_{i-1} - x_i)\,, \tag{6.8}$$

where x_i^*, called an *intermediate point,* is any point in the interval $[x_1, x_{i+1}]$ and consider the following limit,

$$\lim_P S(P; f)\,,$$

where the limit is taken over all partitions P and all sets of intermediate points such that

$$||P|| = \max_{0 \le i \le n-1} (x_{i+1} - x_i) \to 0 .$$

If this limit exists, then f is said to be *Riemann-integrable*, the value of the limit is *the definite integral of* $f(x)$ *over the interval* $[a, b]$ and we use the notation

$$I[f] = \int_a^b f(x)\, dx$$

to denote this limit.

For most functions we need only consider either of two special Riemann sums, called the upper and lower sums and defined as follows. Let

$$m_i = \min_{x_i \le x \le x_{i+1}} f(x) \quad \text{and} \quad M_i = \max_{x_i \le x \le x_{i+1}} f(x) .$$

We define the *upper sum* and *lower sum* by

$$U(P; f) = \sum_{i=0}^{n-1} M_i (x_{i+1} - x_i)$$

and

$$L(P; f) = \sum_{i=0}^{n-1} m_i (x_{i+1} - x_i) .$$

It is clear that for any partition P we have

$$L(P; f) \le I[f] \le U(P; f) .$$

This is illustrated by Figure 6.2 below.

The reader can also observe, by adding an additional point to the partition, that if $\overline{P}$ is a *refinement* of P, that is, $P \subset \overline{P}$ then

$$L(P; f) \le L(\overline{P}; f) \le I[f] \le U(\overline{P}; f) \le U(P; f) .$$

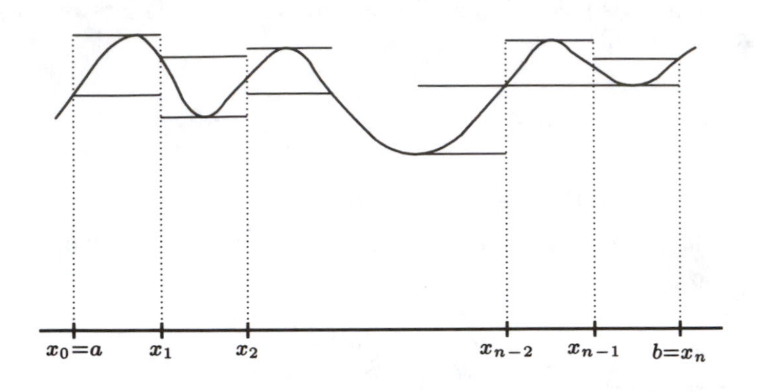

Figure 6.2

Problem 6.3: Sketch an example which illustrates the last set of inequalities.

Our situation is now similar to having partial sums for an alternate series considered in Section 2.1. In particular, if

$$||P|| = \max_i (x_{i+1} - x_i)$$

then we have

Theorem 6.1: *If f is continuous on $[a, b]$ then*

$$\lim_{||P|| \to 0} L(P; f) = \lim_{||P|| \to 0} U(P; f) = I[f].$$

In fact, this theorem can be strengthened to include f which are bounded and piecewise continuous.

Since the Riemann sums must converge for all partitions we often use the following equally spaced partition points:

$$x_k = a + \frac{(b-a)k}{n}, \quad k = 0, 1, \ldots, n. \tag{6.9}$$

Since $x_{k+1} - x_k = \frac{b-a}{n}$ we get

$$U(P_n; f) = \frac{b-a}{n} \sum_{i=0}^{n-1} M_k$$

and

$$L(P_n; f) = \frac{b-a}{n} \sum_{i=0}^{n-1} m_k \,.$$

For example, if

$$I = \int_0^1 \frac{dx}{1+x^2}$$

we see that, since the integrand $f(x) = \frac{1}{1+x^2}$ is decreasing on the interval $[0, 1]$,

$$M_k = f(x_{k-1}) \quad \text{and} \quad m_k = f(x_k) \,.$$

Thus we have

$$U(P_n; f) = \frac{1}{n} \sum_{i=0}^{n-1} \frac{1}{1+x_i^2}$$

and

$$L(P_n; f) = \frac{1}{n} \sum_{i=1}^{n} \frac{1}{1+x_i^2} \,.$$

If we take $n = 1000$, we find that

$$U(P_{1000}; f) \approx 0.7861484734 \quad \text{and} \quad L(P_{1000}; f) \approx 0.7851474724 \,,$$

whereas the correct value is $\arctan(1) = 0.7853981684$.

Problem 6.4: Justify carefully the results of the last paragraph.

Problem 6.5: Suppose in the last section we choose $n = 2$ or $n = 3$. Is $L(P_2; f) \leq L(P_3; f) \leq I[f] \leq U(P_3; f) \leq U(P_2; f)$ true? What about a general function $f(x)$?

In the remaining sections of this chapter we shall discuss other methods that can be used to obtain better results, but they are all based on the theorem presented above.

6.3 Numerical Quadrature

The purpose of this section is to introduce the concept of numerical quadrature to approximate the definite integral of the function f on $[a, b]$. We will obtain the quadrature rules (with errors) usually called the rectangular rule, the trapezoid rule and Simpson's rule.

We begin with a function f on $[a, b]$ for which the definite integral

$$I[f] = \int_a^b f(x)\, dx \tag{6.10}$$

makes sense. We assume $a = x_0 < x_1 < \cdots < x_n = b$ is a partition of $[a, b]$ with corresponding values $f(x_0), f(x_1), \ldots, f(x_n)$ of f. By *numerical quadrature* we mean a rule or formula of the type

$$I_n[f] = \sum_{i=0}^{n} a_i f(x_i)\,. \tag{6.11}$$

The hope is that we can choose the numbers $a_0, a_1, \ldots, a_n$ so that $I_n[f]$ is a good approximation to $I[f]$.

We begin by deriving two important identities:

$$\int_{x_0}^{x_1} f(x)\, dx = h\frac{f(x_0) + f(x_1)}{2} - \frac{1}{12}h^3 f''(\xi_1) \tag{6.12}$$

$$\int_{x_0}^{x_1} f(x)\, dx = hf(y_0) + \frac{1}{24}h^3 f''(\xi_2) \tag{6.13}$$

where $x_0 < \xi_1 < x_1$, $x_0 < \xi_2 < x_1$, $x_1 - x_0 = h$, $y_0 = \frac{x_0+x_1}{2}$ and we assume that f has a continuous second derivative on $[x_0, x_1]$.

Identity (6.13) is obtained by integrating both sides of

$$f(x) = f(y_0) + (x - y_0)f'(y_0) + \frac{(x - y_0)^2}{2} f''(\xi(x))$$

over the interval $[a, b]$ and using Theorem 2.8 of Section 2.2. Thus,

$$\int_{x_0}^{x_1} f(x)\,dx = f(y_0)\int_{x_0}^{x_1} dx + f'(y_0)\int_{x_0}^{x_1}(x - y_0)\,dx + \frac{f''(\xi)}{2}\int_{x_0}^{x_1}(x - y_0)^2\,dx$$

and (6.13) is immediately obtained. Identity (6.12) is obtained by averaging the series

$$f(x_0) = f(y_0) + (x_0 - y_0)f(y_0) + \frac{(x_0 - y_0)^2}{2} f''(\mu_1)$$

and

$$f(x_1) = f(y_0) + (x_1 - y_0)f'(y_0) + \frac{(x_1 - y_0)^2}{2} f''(\mu_2)$$

to obtain

$$\frac{f(x_0) + f(x_1)}{2} = f(y_0) + \frac{1}{2}\left(\frac{h^2}{8}\right) 2f''(\mu_3)\,.$$

This last result follows since $x_0 - y_0 = -\frac{h}{2}$ and $x_1 - y = \frac{h}{2}$ and the fact that $f''(x)$ is continuous allows us to use Theorem 2.3 of Section 2.2 which implies that there exists μ_3 with $\mu_1 < \mu_3 < \mu_2$ so that the average value of $f''(\mu_2)$ and $f''(\mu_2)$ is obtained. Finally, from (6.13) and the last equality we have

$$\int_{x_0}^{x_1} f(x)\,dx = h\left(\frac{f(x_0)+f(x_1)}{2} - \frac{h^2}{8} f''(\mu_3)\right) + \frac{1}{24} h^3 f''(\xi_2)\,.$$

Identity (6.12) now follows by Theorem 2.3 of Section 2.2, where ξ_1 is between μ_3 and μ_2 and hence $x_0 < \xi_1 < x_1$, since $-\frac{1}{8} + \frac{1}{24} = -\frac{1}{12}$.

Problem 6.6: Find ξ_1 and ξ_2 in (6.12) and (6.13), respectively, when $x_0 = 1$, $x_1 = 1$ and $f(x) = \frac{1}{1+x^2}$.

In fact, if f has enough derivatives we may obtain

$$\begin{aligned} \int_{x_0}^{x_1} f(x)\,dx = h\frac{f(x_0)+f(x_1)}{2} - \frac{1}{12} h^3 f''(y_0) \\ - \frac{1}{480} h^5 f^{(iv)}(y_0) + O(h^6) \end{aligned} \tag{6.14}$$

and

$$\begin{aligned} \int_{x_0}^{x_1} f(x)\,dx = hf(y_0) + \frac{1}{24} h^3 f''(y_0) \\ - \frac{1}{1920} h^5 f^{(iv)}(y_0) + O(h^6) \end{aligned} \tag{6.15}$$

where $y_0 = \frac{x_1+x_0}{2}$.

These results are obtained as the results in (6.12) and (6.13) were obtained. Identity (6.15) is found by expanding $f(x)$ in a Taylor series about y_0, integrating both sides over $[x_0, x_1]$ and noting that

$$\int_{x_0}^{x_1} (x-y_0)^m\,dx = \begin{cases} \dfrac{h^{m+1}}{2^m(m+1)} & \text{if } m \text{ is even} \\ 0 & \text{if } m \text{ is odd.} \end{cases}$$

This last result is obtained by a change of variable and recalling what happens to even and odd functions when integrated between symmetric limits. Since $y_0 = \frac{x_0+x_1}{2}$, we let $u = x - y_0$ and note that when $x = x_0$, then $u = \frac{x_0-x_1}{2} = -\frac{h}{2}$ and when $x = x_1$, then $u = \frac{x_1-x_0}{2} = \frac{h}{2}$. Thus

$$\begin{aligned}
\int_{x_0}^{x_1} (x - y_0)^m \, dx &= \int_{-h/2}^{h/2} u^m \, du \\
&= \begin{cases} 2\int_0^{h/2} u^m \, du & \text{if } m \text{ is even} \\ 0 & \text{if } m \text{ is odd} \end{cases} \\
&= \begin{cases} \frac{h^{m+1}}{2^m(m+1)} & \text{if } m \text{ is even} \\ 0 & \text{if } m \text{ is odd}, \end{cases}
\end{aligned}$$

as claimed.

Problem 6.7: Using the results for (6.13) as a hint, give a different expression for (6.15) incorporating the value ξ_3. Prove this result.

Identity (6.14) is found by expanding both $f(x_0)$ and $f(x_1)$ in a Taylor series about y_0, averaging the two series, solving for $hf(y_0)$ and substituting this value into the right hand side of (6.15). Formulas (6.12) and (6.13) are special cases of (6.14) and (6.15).

Let us note that we do get something for nothing by adding one third of identity (6.14) to two thirds of identity (6.15) to obtain

$$\begin{aligned}
\int_{x_0}^{x_1} f(x)\, dx = h\left(\frac{f(x_0) + 4f(y_0) + f(x_1)}{6}\right) \\
- \frac{h^5}{2880} f^{(iv)}(y_0) + O(h^6).
\end{aligned}$$

This can be put into more familiar form associated with Simpson's Rule if we replace h by $2h$, $y_0 = x_0 + \frac{h}{2}$ with $x_1 = x_0 + h$ and $x_1 = y_0 + h$ with $x_2 = x_0 + 2h$. We obtain

$$\int_{x_0}^{x_2} f(x)\, dx = h\left(\frac{f(x_0) + 4f(x_1) + f(x_2)}{3}\right) - \frac{h^5}{90} f^{(iv)}(\xi) \tag{6.16}$$

for $x_0 < \xi < x_2$.

Problem 6.8: Find ξ in (6.16) when $x_0 = 0$, $x_2 = 2$ and $f(x) = \frac{1}{1+x^2}$.

The error terms such as in (6.12), (6.13) or (6.16) show that for some functions $f(x)$, some quadrature rules will involve no error. As an example, if we let $f(x) = x^2$ and $x_0 = 1$ we have the exact answer

$$\int_1^{1+h} x^2\, dx = \frac{1}{3}x^3\bigg|_1^{1+h} = \frac{1}{3}((1+h)^3 - 1) = h + h^2 + h^3 .$$

Applying the right hand size of (6.12) we have

$$\frac{h}{2}(1^2 + (1+h)^2) - \frac{1}{12}h^3(2) = h + h^2 + \frac{h^3}{3}$$

while applying the right hand side of (6.13) we have

$$h\left(1 + \frac{h}{2}\right)^2 + \frac{1}{24}h^2(2) = h + h^2 + \frac{h^3}{3} .$$

The right hand side of (6.16), with $\frac{h}{2}$ instead of h is

$$\frac{h}{6}\left(1^2 + 4\left(1 + \frac{h}{2}\right)^2 + (1+h)^2\right) = h + h^2 + \frac{h^2}{3} .$$

The moral is that formulas

$$T_1[f] = h\,\frac{f(x_0) + f(x_1)}{2} \tag{6.17}$$

and

$$R_1[f] = hf\left(\frac{x_0 + x_1}{2}\right) \tag{6.18}$$

are not exact for quadratic functions, although they are exact for linear functions. Similarly, formula

$$S_2[f] = h\,\frac{f(x_0) + 4f(x_1) + f(x_2)}{3} \tag{6.19}$$

is exact for a cubic polynomial. By exact we mean these formulas give the same value as $I[F]$ in (6.10). In fact, these formulas can be derived from these properties of exactness (see Exercises 6.8 and 6.15).

Problem 6.9: Show that (6.10) and (6.19) agree for the functions $f_1(x) = 1$, $f_2(x) = x$, $f_3(x) = x^2$, $f_4(x) = x^3$. Show that the same result holds for $f(x) = c_3x^3 + c_2x^2 + c_1x + c_0$ by using the linearity of S_2 and I.

We will now generalize the formulas (6.17), (6.18), and (6.19) to obtain approximations to the definite integral of f on $[a, b]$. The approximations we will obtain are *composition formulas* formed by summing the respective formulas (6.17), (6.18), and (6.19). We will assume that the points of our partition are equally spaced, so that $x_{i+1} = x_i + h$, to obtain a more symmetrical formula, but we could assume that they are not equally spaced.

Using (6.10) and (6.12) we have

$$\begin{aligned} I[f] = \int_a^b f(x)\,dx &= \sum_{i=0}^{n-1} \int_{x_i}^{x_{i+1}} f(x)\,dx \\ &= \frac{h}{2} \sum_{i=0}^{n-1} (f(x_i) + f(x_{i+1})) \\ &= \frac{1}{12} h^3 \sum_{i=0}^{n-1} f''(\xi_i) \end{aligned}$$

where $x_i < \xi_i < x_{i+1}$ for $i = 0, 1, \ldots, n-1$. The next figure illustrates this result. The area in the trapezoids under the dotted lines are given by the first sum and in (6.20a) below. The second sum or (6.20b) below is the error, the net area between the two curves. The first panel which is the picture between $x = x_0$ and $x = x_1$ similarly illustrates (6.12).

The formula

$$T[f] = \frac{h}{2}(f_0 + 2f_1 + 2f_2 + \cdots + 2f_{n-1} + f_n) \tag{6.20a}$$

is called the *Trapezoid Rule.* We will soon see that the corresponding error expression $-\frac{1}{12}h^3 \sum_{i=0}^{n-1} f''(\xi_i)$ becomes

$$E_T[f] = -\frac{(b-a)h^2}{12} f''(\xi) \tag{6.20b}$$

where ξ is such that $a < \xi < b$.

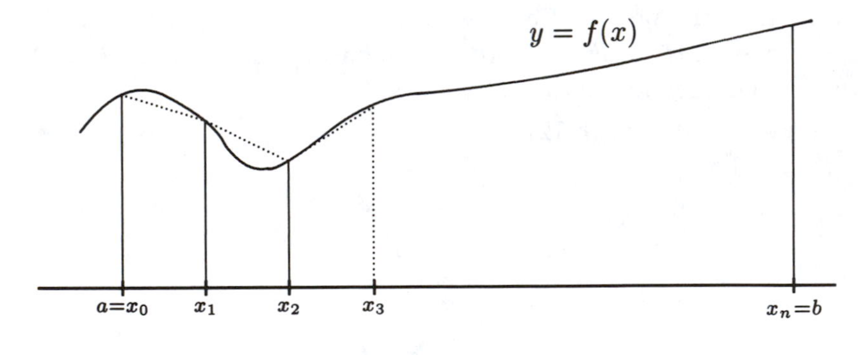

Figure 6.3

It is important to understand the error term. From (6.12), the error expression

$$-\frac{1}{12}h^3 f''(\xi_i)$$

is the *local error* of the trapezoid rule. Note the *global error* for the trapezoid rule is

$$-\frac{h^3}{12}\sum_{i=0}^{n-1} f''(\xi_i),$$

where $x_i < \xi_i < x_{i+1}$ for $i = 0, 1, \ldots, n-1$. If we assume that f has a continuous second derivative on $[a, b]$, then we can produce the result of (6.20b). Note that, since $h = \frac{b-a}{n}$, we have

$$-\frac{h^3}{12}\sum_{i=0}^{n-1} f''(\xi_i) = -\frac{h^2}{12}(b-a)\frac{1}{n}\sum_{i=0}^{n-1} f''(\xi_i). \tag{6.21}$$

If we note that averages are always between the smallest and largest term in the set being averaged, we see that

$$\min_{a\le x\le b} f''(x) \le \frac{1}{n}\sum_{i=0}^{n-1} f''(\xi_i) \le \max_{a\le x\le b} f''(x).$$

Since f'' is assumed to be continuous we see that, by Theorem 2.2 of Chapter 2, f'' actually assumes the values $\min_{a\le x\le b} f''(x)$ and $\max_{a\le x\le b} f''(x)$, and so, by the Intermediate Value Theorem (Theorem 2.3 of Chapter 2), it takes on all values between these two numbers. Thus, there exists a number ξ in the interval (a, b) so that

$$\frac{1}{n}\sum_{i=0}^{n-1} f''(\xi_i) = f''(\xi). \tag{6.22}$$

If we combine (6.21) and (6.22) we get (6.20b).

The moral is that we lose one order of h, as for example in (6.21), by going from the local error to the global error. Intuitively the summation process yields nh which is equal to the constant $b-a$.

Similarly from (6.19), if $2mh = (b-a)$, we have *Simpson's Rule*

$$\begin{aligned} S[f] = \frac{h}{3}(f_0 + 4f_1 + 2f_2 + 4f_3 + 2f_4 + \cdots \\ + 2f_{2m-2} + 4f_{2m-1} + f_{2m}) \end{aligned} \tag{6.23a}$$

with corresponding *global error* in Simpson's rule

(6.23b) $$E_S[f] = -\frac{(b-a)h^4}{180} f^{(iv)}(\xi) .$$

Problem 6.10: Derive (6.23b). You may want to use the argument which justified (6.20b).

The errors (6.20b) or (6.23b) are examples of truncation errors. We recall that there is a second type of error associated with quadrature rules such as (6.20a) or (6.23a), namely round-off error, which is the machine error in adding many numbers together. It can be shown that this error is also $O(h^m)$.

It is instructive to picture the truncation errors vs. round-off errors. The graph below illustrates the general situation and is independent of the problem, the rule or the computer hardware. The x-axis is associated with the number of subdivisions of the interval. The y-axis represents the error. The straight line $y = y_s(x)$ is the single precision round-off error function. The straight line $y = y_d(x)$ is the double precision round-off error curve. The straight line $y = y_t(x)$ is the truncation error curve. The reader should understand why the functions $y_s(x)$, $y_d(x)$ and $y_t(x)$ are linear functions since the associated error terms are powers of h. For effect we have pictured these curves as straight lines and the points on the curve.

The moral is that the trade off between truncation error and round-off error leads to a best number of partitions which depends on the problem, the rule and the precision of the computer hardware. For relatively large $h > 0$, reducing h decreases the truncation error as the picture indicates but has little effect on the round-off error which is negligible. Conversely for very small $h > 0$, decreasing h decreases the truncation error very little in comparison with the round-off error which is increasing.

Our final idea of this section is to see that if our error is $O(h^m)$, then a Richardson extrapolation procedure gives us "something for nothing." Hence, as we see below if we put in twice as many partition points by replacing h with $\frac{h}{2}$ the new global error $E_{h/2}$ in the

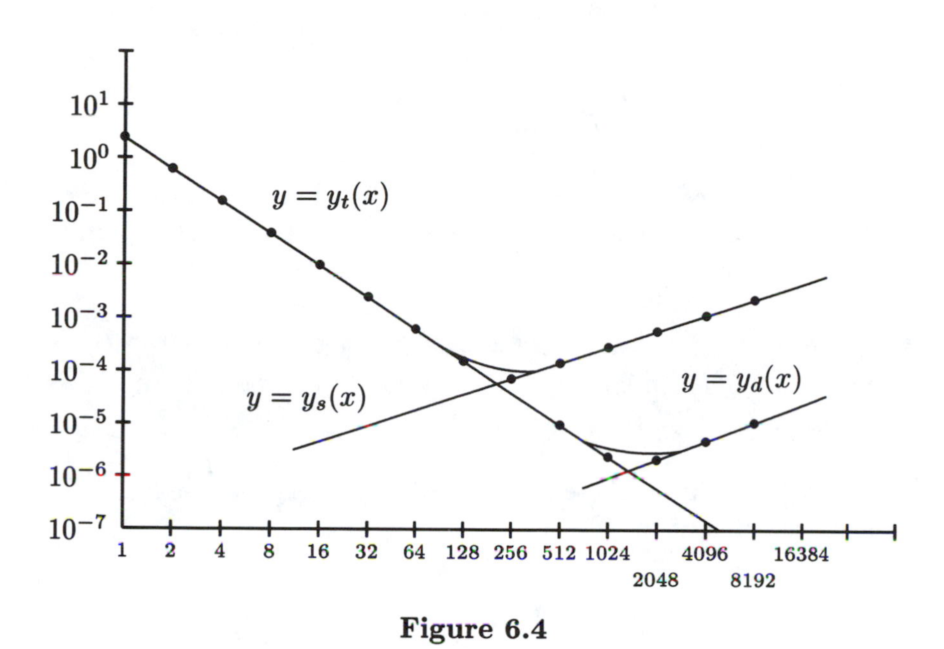

Figure 6.4

trapezoid rule would be approximately one quarter of the original error E_h while the new global error in Simpson's rule would be approximately one sixteenth of the original error. In fact, we can do better by the process of Richardson extrapolation. Let

$$E_h = I[f] - T_k[f] = c_1 h^2 + c_2 h^3 + \cdots$$

be the error as a function of h. Then

$$E_{h/2} = c_1 \frac{h^2}{4} + c_2 \frac{h^3}{8} + \cdots .$$

If we define

$$T_h^R[f] = \frac{4T_{h/2}[f] - T_h[f]}{3} \tag{6.24a}$$

then

$$I[f] - T_h^R[f] = O(h^3)\,. \tag{6.24b}$$

Thus if we replace h by $\frac{h}{2}$, instead of nearly obtaining a quarter of the error we may use (6.24) to obtain results which are $O(h^3)$. Truly, something for nothing.

Problem 6.11: If $E_h = c_1h^2 + c_2h^4 + \cdots$ give a Richardson extrapolation $\overline{T}_h^R[f]$ similar to (6.24a). What is the value of m if this new error is $O(h^m)$?

As an example of these results, let $f(x) = \frac{1}{x^2+1}$ so that we wish to find the value of the integral

$$\int_0^1 \frac{\mathrm{dx}}{x^2+1}\,.$$

The reader will recall that this integral is equal to $\arctan(1) = 0.785398163$. In the table below we give the results of applying the trapezoid rule and Simpson's rule to the calculation of this integral for three step sizes. Each step is one half of the previous step so that the error decreases, in accordance with (6.20b) and (6.23b), should be observable. Also, we have included the computations based on the use of Richardson extrapolation. The reader may observe that the errors in the trapezoid error column decrease by approximately one-quarter as $\frac{h}{2}$ replaces h. We also note that the Richardson values are much more accurate than the trapezoid values although our expectation that the errors in the Richardson trapezoid error column, or equivalently in the Simpson error column, will decrease by approximately one-eighth is less accurate.

Problem 6.12: Do the values of the trapezoid and Simpson errors in Table 6.1, below, seem reasonable? Explain. Explain in what sense the Richardson errors are reasonable and in what sense they are not.

Correct value for $h = 0.125$, $h = 0.0625$, and $h = 0.03125$ below is 0.78539816339745.

h	Trapezoid Value	Error	Richardson Value	Error
0.125	0.78474712362277	6.5103977468E-04	—	—
0.0625	0.78523540301035	1.6276039710E-04	0.78539816280621	5.9124E-10
0.03125	0.78553574739374	4.069010371E-05	0.78539816338820	9.25E-12

h	Simpson Value	Error	Richardson Value	Error
0.125	0.78539816339745	3.778277E-08	—	—
0.0625	0.78539816280621	5.9124E-10	0.78539816328564	1.1181E-10
0.03125	0.78539816338821	9.24E-12	0.78539816339658	8.7E-13

Table 6.1

6.4 Newton-Cotes Formulas

For equally spaced points $a = x_0 < x_1 < \cdots < x_n = b$ where $x_1 = x_0 + ih$ $(i = 0, 1, 2, \ldots, n)$ we obtain a large class of quadrature rules from the Lagrange Interpolation Formula. We have already seen the first two of these rules, the trapezoid rule and Simpson's rule, which are given in the previous section. Thus, this brief section is for understanding and completeness. We will summarize the results in the next paragraph and then try to explain what these results really say to the reader.

We recall the Lagrange Interpolation Formula

$$f(x) = \sum_{j=0}^{n} \prod_{\substack{i=0 \\ i \neq j}}^{n} \frac{x - x_i}{x_j - x_i} f(x_j) - \frac{1}{(n+1)!} f^{(n+1)}(\xi) \prod_{j=0}^{n} (x - x_j) . \tag{6.25}$$

Integrating both sides over the interval $[a, b]$ we have after several steps

$$\int_a^b f(x)\, dx = \sum_{k=0}^{n} a_k f(x_k) + E_{n+1}(f) \tag{6.26}$$

where

$$a_k = \frac{(-1)^{n-k} h}{k!(n-k)!} \int_0^n t(t-1) \cdots (t-i-1) \cdot (t-i+1) \cdots (t-n)\, dt \tag{6.27}$$

and

$$E_{n+1}(f) = \frac{1}{n+1} \int_a^b f^{(n+1)}(\xi(x)) \prod_{j=0}^{n} (x - x_j)\, dx .$$

For completeness, we note that, a sophisticated use of an integral mean value theorem yields the result that, if n is even and $f^{(n+2)}$ is continuous in $[a, b]$ then

$$E_{n+1}(f) = \frac{h^{n+3} f^{(n+2)}(\xi)}{(n+2)!} \int_0^n t^2 (t-1)(t-2) \cdots (t-n)\, dt \tag{6.28a}$$

while if n is odd and $f^{(n+1)}$ is continuous on $[a, b]$ then

$$E_{n+1}(f) = \frac{h^{n+2} f^{(n+1)}(\xi)}{(n+1)!} \int_0^n t(t-1) \cdots (t-n)\, dt . \tag{6.28b}$$

These results are given in more complete form in Theorem 6.2, below.

The application of (6.25)–(6.27) with $n = 1$ yields the trapezoid rule of the previous section. As an illustration, we will now apply these results to the case $n = 2$ to obtain Simpson's rule, which was obtained in (6.16) of the previous section using an alternate method. Thus, we begin with

$$\begin{aligned} f(x) = {} & \frac{(x - x_1)(x - x_2)}{(x_0 - x_1)(x_0 - x_2)} f(x_0) \\ & + \frac{(x - x_0)(x - x_2)}{(x_1 - x_0)(x_1 - x_2)} f(x_1) \\ & + \frac{(x - x_0)(x - x_1)}{(x_2 - x_0)(x_2 - x_1)} f(x_2) \\ & + \frac{1}{3!} f'''(\xi)(x - x_0)(x - x_1)(x - x_2)\,. \end{aligned}$$

Integrating both sides of this equality and using the change of variable $x = x_0 + th$ which implies that $x_k - x_j = x_0 + hk - (x_0 + jh) = (k - j)h$ and $x = x_k + \frac{t-k}{h}$ for $j, k = 0, 1, \ldots, n$, we obtain

$$\begin{aligned} \int_a^b f(x)\, dx = {} & hf(x_0) \int_0^2 \frac{(t-1)(t-2)}{(-1)(-2)}\, dt \\ & + hf(x_1) \int_0^2 \frac{(t-0)(t-2)}{(1)(-1)}\, dt \\ & + hf(x_2) \int_0^2 \frac{(t-0)(t-1)}{(2)(1)}\, dt \\ & + \frac{h^4}{6} \int_0^2 f'''(\xi(x)) t(t-1)(t-2)\, dt \\ = {} & \frac{h}{2} f(x_0) \left(\frac{2}{3}\right) - hf(x_1) \left(-\frac{4}{3}\right) \\ & + \frac{h}{2} f(x_2) \left(\frac{2}{3}\right) + E_2(f) \\ = {} & \frac{h}{3} \left(f(x_0) + 4f(x_1) + f(x_2)\right) + E_2(f)\,. \end{aligned}$$

We point out that the calculation

$$\begin{aligned}
E_2(f) &= \frac{h^4}{6}\int_0^2 f'''(\xi(x))t(t-1)(t-2)\ dt \\
&= \frac{h^4 f'''(\eta)}{6}\int_0^2 t(t-1)(t-2)\ dt \\
&= 0
\end{aligned}$$

for some η between a and b, is not correct. The problem is that $g(t) = t(t-1)(t-2)$ changes sign on $[0, 2]$ so that Theorem 2.8 of Chapter 2 is not applicable. The actual calculation of $E_2(f)$, producing the result in (6.27), can be obtained by applying the Mean Value Theorem to the intervals $[0, 1]$ and $[1, 2]$ separately (see Exercise 6.21).

Finally we note, that if $n = 1$, the error for the trapezoid rule causes no problem since $y(t) = t(t-1)$ does not change sign on $(0, 1)$ so that our change of variables yields

$$\begin{aligned}
E_1(f) &= \frac{1}{2!}\int_a^b f''(\xi(x))(x-x_0)(x-x_1)\ dx \\
&= \frac{1}{2}\int_0^1 f''(\xi)(th)(t-1)h(h\ dt) \\
&= \frac{h^3}{2}f''(\eta)\int_0^1 t(t-1)\ dt \\
&= -\frac{h^3}{12}f''(\eta)
\end{aligned}$$

for some value of η, $a < \eta < b$. This result agrees with Formula (6.12) of the previous section.

Problem 6.13: Work out the complete details when $n = 1$ for the trapezoid rule starting with (6.25).

Finally, for completeness we note that the above formulas or identities given in (6.26)–(6.28) are called *Newton-Cotes Closed Formulas* or Identities as they include the endpoints $a = x_0$ and $b = x_n$. If we begin with the interpolation formula which interpolates the interior points $x_1 < x_2 < \cdots < x_{n-1}$, the resulting formulas are called *Newton-Cotes Open Formulas.*

We close this section with two final remarks. Although, as was mentioned above, it is a difficult task to compute the error estimates it can be done with some extra effort and one obtains the following results which we have given above.

Theorem 6.2: *If n is even and $f(x)$ is $n+2$ times continuously differentialbe on $[a,b]$, then*

$$E_n(f) = C_n h^{n+3} f^{(n+2)}(\eta)\,,$$

for some $\eta \in (a,b)$ with

$$C_n = \frac{1}{(n+2)!} \int_0^n t^2(t-1)\cdots(t-n)\,dt\,.$$

If n is odd and $f(x)$ is $n+1$ times continously differentiable on $[a,b]$, then

$$E_n(f) = C_n h^{n+2} f^{(n+1)}(\eta)\,,$$

for some $\eta \in (a,b)$, with

$$C_n = \frac{1}{(n+1)!} \int_0^n t(t-1)\cdots(t-n)\,dt\,.$$

It should be noted that the Newton-Cotes formulas with an even index n gain an extra degree of precision as compared with those of odd index $n+1$. For example, both $n=2$ and $n=3$ have the same error estimate, namely $O(h^5)$.

It might be assumed from the error estimates of the theorem that if we increase n, then we shall get better results. Unfortunately this is not always the case. The following table gives the results for our standard integral, namely

$$\int_0^1 \frac{\mathrm{dx}}{1+x^2},$$

whose value is $\frac{\pi}{4} \approx 0.7853981684$. We have chosen to list the values only for the Newton-Cotes with even n, since they have the same order of error as the following odd values.

n	computed value	error
2	0.783333325	-2.0648434E-03
4	0.785529405	1.312366E-04
6	0.7853927113	-5.4571E-06
8	0.7881783362	2.7801678E-03
10	0.409754463	-0.3756537054

Table 6.2

Note that the problems begin with $n=8$ and the value for $n=10$ is quite far off. The reason for this is simple: the formulas for $n=8$ and $n=10$ contain negative weights. This allows for cancellation, which is what happens with $n=10$ to a very large degree. Also as n increases one would suspect, based on the formula for the weights (6.26), that the denominators of these rational numbers will increase.

Thus, with negative weights and large denominators, the calculations become somewhat delicate and incorrect computer results are very likely.

The moral to the story is that we do not want to use Newton-Cotes formulas for very large values of n. Indeed, most people do not use values of $n \geq 8$, even in composite formulas.

6.5 The Romberg Algorithm

In this section we return to the trapezoid rule, which says that we approximate the definite integral of $f(x)$ over the interval $[a, b]$ by the sum

$$T(P_n; f) = h \sum_{i=1}^{n-1} f(a + ih) + \frac{h}{2} (f(a) + f(b)) \,, \tag{6.29}$$

where $h = \frac{b-a}{n}$. We know that, to a certain extent, if n is made larger, then the approximation will be better, as long as we do not exceed the point where round-off error begins to become a big factor. This idea will be combined with extended Richardson extrapolation to produce the triangular array in Table 6.3, below.

To understand and use Romberg methods it is convenient to use $2^n + 1$ points in a partition instead of the $n + 1$ points that are used in P_n in (6.29). This allows us to double the number of panels (or, what is the same thing, halve h) rather easily. It also turns out that there is a nice relation between the $2^n + 1$ points of our new partition and the $2^{n-1} + 1$ points of our old partition that cuts the calculation in half as we need only calculate the value of f at new points while using the previous values of f at the points of the old partition.

Let $R(n+1, 1)$ denote the trapezoid sum for a partition of $2^n + 1$ equally spaced points (we shall explain the notation below). Then, from (6.29), with 2^n replacing n we have

$$
\begin{aligned}
R(n+1,1) &= \frac{b-a}{2^n}\sum_{i=1}^{2^n-1} f\left(a+i\frac{b-a}{2^n}\right) \\
&\qquad + \frac{b-a}{2^{n+1}}(f(a)+f(b)) \\
&= \frac{b-a}{2^n}\sum_{j=1}^{2^{n-1}-1} f\left(a+2j\frac{b-a}{2^n}\right) \\
&\qquad + \sum_{j=1}^{2^n-1} f\left(a+(2j-1)\frac{b-a}{2^n}\right) \\
&\qquad + \frac{b-a}{2^{n+1}}(f(a)+f(b)) \\
&= \frac{1}{2}R(n,1) + \frac{b-a}{2^n} \\
&\qquad \cdot \sum_{j=1}^{2^n-1} f\left(a+(2j-1)\frac{b-a}{2^n}\right).
\end{aligned} \tag{6.30}
$$

Thus to calculate $R(n+1,1)$ after we have $R(n,1)$ we need only calculate the function values at the new partition points, that is, the points midway between x_i and x_{i+1} of the old partition. We do not need to recalculate f at the old partition points x_i.

We now tie in another idea from before, namely Richardson extrapolation. If we begin with formula (6.10) of Section 6.3 and repeat our discussion of local error and global error we obtain

$$\int_a^b f(x)\,dx = R(n,1) + a_2h^2 + a_4h^4 + \cdots, \tag{6.31}$$

where the a_i depend on f but not on h. If we replace h by $\frac{h}{2}$, this replaces $R(n,1)$ by $R(n+1,1)$ and we get

$$\int_a^b f(x)\,dx = R(n+1,1) + \frac{1}{4}a_2h^2 + \frac{1}{16}a_4h^4 + \cdots. \tag{6.32}$$

Now, as with Richardson extrapolation, we multiply (6.32) by $\frac{4}{3}$ and subtract $\frac{1}{3}$ of (6.31) from it to obtain

$$\int_a^b f(x)\,dx = \frac{4}{3}R(n+1,1) - \frac{1}{3}R(n,1) - \frac{1}{4}a_4h^4 - \cdots. \tag{6.33}$$

Let us call the main term $R(n,2)$, that is,

$$R(n,2) = \frac{4}{3}R(n+1,1) - \frac{1}{3}R(n,1).$$

Here the second index 2 indicates that we have performed one extrapolation on the trapezoid sums $R(n,1)$.

If we repeat this process we obtain, for the mth extrapolation, the following relation

$$R(n+1,m+1) = \frac{4^m R(n+1,m) - R(n,m)}{4^m - 1}, \tag{6.34}$$

where $n = 1, 2, \ldots$ and $m = 1, 2, \ldots, n$. This produces the triangular array

$$\begin{array}{lllll}
R(1,1) & & & & \\
R(2,1) & R(2,2) & & & \\
R(3,1) & R(3,2) & R(3,3) & & \\
\vdots & \vdots & \vdots & & \\
R(n,1) & R(n,2) & R(n,3) & \cdots & R(n,n),
\end{array}$$

Table 6.3

where the first column is produced from the trapezoid rule with the aid of (6.30) and the remaining columns are produced by formula (6.34) from the previous column using the element across from it and the element one up.

There are several important advantages of this method over trapezoid sums. The first advantage is that the second, third, ... columns are obtained with "no" extra effort in that while (6.34) is performed no new values of $f(x)$ are needed. Thus, more accuracy is obtained for the same effort. Often more important is that since we can use larger values of h to obtain a desired accuracy, round-off errors are usually not a factor as they would be if only the trapezoid sums in the first column are used.

After producing this array we usually take $R(n, n)$ as the final approximation once we decide on the value of n. Note, however, that the coefficients a_i, in the analogs of (6.32) for the columns corresponding to $m = 3, 4, \ldots$, depend on f. Thus they could be reasonably large and hence, increasing the values of n and m may only compound the error. To decide whether to accept $R(n+1, n+1)$ as an improvement over $R(n, n)$ we rewrite (6.34) and consider the ratio

$$\frac{R(n+2, n) - R(n+1, n)}{R(n+1, n) - R(n, n)}.$$

If this ratio, when computed, is near to 4^n, then we take $R(n+1, n+1)$ over $R(n, n)$. If it isn't near 4^n, then we stop at $R(n, n)$.

As an example we take our standard example

$$I = \int_0^1 \frac{dx}{x^2+1}$$

with $I = \frac{\pi}{4} \approx 0.7853981684$. If we take $n = 6$, we get the array of numbers in Table 6.4, below. Thus we take as our sequence of approximations:

.75, .783333333, .7855294114, .7853964442 .7853981631, .785391609,

which gives us, with $n = 6$, seven place accuracy. These numbers are the diagonal entries in Table 6.4.

.75					
.775	.78333333				
.7827841175	.7853921566	.7855294114			
.7847471224	.785398124	.7853985218	.7853964442		
.7852354005	.7853981598	.7853981621	.7853981564	.7853981631	
.7853574708	.7853981609	.7853981621	.7853981609	.7853981609	.7853981609

Table 6.4

Problem 6.14: Using the value $\frac{\pi}{4} \approx 0.7853981684$ underline the correct digits in Table 6.4. More precisely, make a table similar to Table 6.4 of errors. Verify by inspection that the errors are reasonable as we move down each row or across each column.

Exercise Set 6

1. Derive formulas (6.6) and (6.7) of Section (6.1).

2. Show that

$$
\begin{aligned}
f'(x_0) &= -\frac{f_2 + 4f_1 - 3f_0}{2h} + O(h^2), \\
f''(x_0) &= \frac{f_2 - 2f_1 + f_0}{h^2} + O(h^2), \quad \text{and} \\
f'''(x_0) &= \frac{f_2 - 2f_1 + 2f_{-1} - f_{-2}}{2h^3} + O(h^2).
\end{aligned}
$$

3. Suppose $x_1 < x_2 < x_3$ and $x_2 - x_1 = h$ and the number α is defined by $x_3 - x_2 = \alpha h$.

   a) Show that $f''(x)$ can be approximated by

   $$\frac{2}{h^2}\left(\frac{f(x_1)}{1+\alpha} - \frac{f(x_2)}{\alpha} + \frac{f(x_3)}{\alpha(\alpha+1)}\right).$$

   (HINT: determine constants A, B and C so that

   $$f''(x) = Af(x_1) + Bf(x_2) + Cf(x_3)$$

   for all polynomials of degree ≤ 2. By linearity one only need consider the polynomials 1, $x - x_2$, and $(x - x_2)^2$.)

   b) Use Taylor series and the results of (a) to show that

   $$f'(x_2) = \frac{f(x_3) - f(x_1)}{x_3 - x_1} + (\alpha - 1)\frac{h}{2}f''(x_2) + O(h^2),$$

   where $x_1 < x_2 < x_3$. Thus the error in approximation $f'(x_2)$ by $\frac{f(x_3)-f(x_1)}{x_3-x_1}$ is $O(h^2)$ if x_2 is the average of x_1 and x_3, but only $O(h)$ elsewise.

4. Let f be an increasing function and suppose we estimate $\int_a^b f(x)\,dx$ by upper and lower sums on equally spaced points. What is the maximum error made?

5. Obtain an error estimate for the midpoint formula, both for equally and nonequally spaced partitions.

6. Let f be defined on $[a, b]$ and let P be a partition of the interval $[a, b]$.

   a) Show that if f is decreasing, then

$$T(P; f) = \frac{1}{2}\big(U(P; f) + L(P; f)\big).$$

   b) Show that for any function f we have

$$L(P; f) \leq T(P; f) \leq U(P; f).$$

7. A function f is said to be *convex* if its graph lies below every chord drawn between two points on the graph. What is the relationship between $L(P; f)$, $U(P; f)$, $T(P; f)$ and $\int_a^b f(x)\, dx$ for such a function?

8. Show that if f is integrable, then

$$\lim_{P} T(P; f) = \int_a^b f(x)\, dx.$$

9. How large must n be in order to compute $\int_0^1 e^{-x^2}\, dx$ to an accuracy of 8 significant figures using the trapezoid rule?

10. Derive the trapezoid rule with error starting with the appropriate Lagrange Interpolation equality.

11. Show that $S(F)$ in (6.19), of Section 6.3, is exact for the function $f(x) = 1$, $f(x) = x$ and $f(x) = x^2$.

12. Show that a Richardson process similar to (6.24), of Section 6.3, applied to Simpson's rule, such that

$$S_h^R[f] = \frac{16S_{h/2}[f] - S_h[f]}{15}$$

yields an error of $O(h^5)$.

13. How large must n be in order to calculate $\int_0^1 e^{-x^2}\,dx$ to an accuracy of 8 significant figures using Simpson's $\frac{1}{3}$ rule?

14. Note that Simpson's rule is exact for cubic polynomials. (Why?) By the remarks in Section 6.4 we see that Simpson's rule is obtained by fitting a quadratic through three equally spaced points. This seems to imply that the area under any cubic polynomial from $x = a$ to $x = b$ is exactly the same as the area under the parabola that intersects the cubic at $x = a$, $x = b$ and also at $x = \frac{a+b}{2}$. Prove this.

15. For $n \geq 2$ show that there exist real numbers $\alpha_1, \ldots, \alpha_n$ such that $\alpha_1 + \cdots + \alpha_n = 1$ and

$$R(n,n) = \alpha_1 R(1,1) + \alpha_2 R(2,1) + \cdots + \alpha_n R(n,1)\,.$$

16. Show that the second column in the Romberg algorithm, $R(n,2)$, is really just Simpson's rule.

17. Determine α, β and γ so that the rule of the form

$$\int_a^b f(x)\,dx = \alpha f\left(\frac{3a+b}{4}\right) + \beta f\left(\frac{a+b}{2}\right) + \gamma f\left(\frac{a+3b}{4}\right)$$

is exact for polynomials of degree ≤ 2.

(HINT: To make the calculations easier make a change of variable so that the interval of integration is $[0,1]$.)

18. Determine values of the constants $\alpha_0, \ldots, \alpha_n$ so that the rule

$$\int_a^b f(x)\,dx = \sum_{j=0}^{n} \alpha_j f\left(a + \frac{j(b-a)}{n}\right)$$

is exact for polynomials of degree $\leq n$. Do the calculations for $n = 1$, 2, and 3.

(HINT: Since polynomials are linear combinations of powers and integrals are linear we need only determine the constants $\alpha_0, \ldots, \alpha_n$ so that the relationship is exact for the powers $1, x, \ldots, x^n$.)

Note that for $n = 1$ we get the coefficients for the trapezoid rule and that for $n = 2$ we get Simpson's rule.

19. Find values of $\alpha_1, \ldots, \alpha_n$, $\xi_1, \ldots, \xi_n$, where $\xi_1, \ldots, \xi_n$ lie in the interval $[a, b]$ so that the quadrature rule

$$\int_a^b f(x)\,dx = \sum_{i=1}^{n} \alpha_i f(\xi_i)$$

is exact for polynomials of degree $\leq 2n - 1$. Do the calculations for $n = 1$ and 2. This is called Gaussian quadrature and is discussed in detail in Chapter 9.

(HINT: By a change of variable we need only consider the integral

$$\int_{-1}^{1} f(x)\,dx$$

and make the formula exact for polynomials of degree $\leq 2n - 1$. As above make the formula exact for the polynomials $1, x, \ldots, x^{2n-1}$ and use linearity. Then translate the formula back to the interval $[a, b]$. The points ξ_i, $i = 1, 2, \ldots, n$ are the roots of a certain class of orthogonal polynomials. See also Section 9.2)

20. Derive the identities (6.14) and (6.15) in Section 6.3.

21. Complete the derivation of the error term $E_2(f)$ for Simpson's Rule as in formula (6.28a), with $n = 2$, of Section 6.4.

Computational Problems

22. Calculate $\frac{\cos(h)-1}{h}$ for $h = 0.01 \cdot 2^{-n}$ for $n = 0, 1, \ldots, 20$. Explain your results.

23. Calculate the following integrals using the trapezoid algorithm.

    a) $\displaystyle\int_0^1 \arctan x \, dx$

    b) $\displaystyle\int_1^2 \cos(\log x) \, dx$

    c) $\displaystyle\int_0^1 \frac{dx}{e^x + e^{-x}}.$

    In each case compare with the correct value. Use enough points to guarantee 6 place accuracy.

24. Use Simpson's $\frac{1}{3}$ algorithm to calculate the following integrals.

    a) $\displaystyle\int_0^1 \arctan x \, dx$

    b) $\displaystyle\int_1^2 \cos(\log x) \, dx$

    c) $\displaystyle\int_0^1 \frac{dx}{e^x + e^{-x}}.$

    Compare with the correct values. Use enough points to guarantee 6 place accuracy.

25. Recall from calculus that the length of a curve is

$$\int_a^b \sqrt{1 + f'(x)^2}\, dx$$

where $y = f(x)$ is the curve in question on the interval $[a, b]$. Use Simpson's rule to calculate the following arclengths. Use a step size that guarantees 6 place accuracy.

   a) The length of the curve $x^2 + y^2 = 1$ and compare with the correct value.

   b) The length of the curve $x^2 + 4y^2 = 1$.

26. Use the Romberg algorithm to its greatest accuracy to calculate the following integrals.

   a) $\displaystyle\int_0^1 \arctan x\, dx$

   b) $\displaystyle\int_1^2 \cos(\log x)\, dx$

   c) $\displaystyle\int_0^1 \frac{dx}{e^x + e^{-x}}$.

   In each case compare with the correct value as well as the values obtained from the previous numerical integration algorithms.

27. The Bessel function of order n, (see also Exercise 2.7 of Chapter 2) can be shown to be equal to the integral

$$J_n(x) = \frac{1}{\pi} \int_0^\pi \cos(x \sin\theta - n\theta)\, d\theta\,.$$

   Use the Romberg algorithm to its greatest accuracy to calculate $J_n(1)$ for $n = 0, 1, 2$.

28. Use Simpson's rule with $h = 0.001$ to calculate the following integrals.

    a) Define

    $$f(x) = \begin{cases} x^2 + 1\,, & 0 \le x \le 3 \\ 2x + 4\,, & 3 \le x \le 4 \end{cases}$$

    and calculate

    $$\int_0^4 f(x)\, dx\,.$$

    b) Calculate

    $$\int_{-1}^3 |e^x - 1|\, dx\,.$$

    In both cases compare the exact values.
    (HINT: Recall that

    $$\int_a^b f(x)\, dx = \int_a^c f(x)\, dx + \int_c^b f(x)\, dx$$

    for any c with $a \le c \le b$.)

7 *Numerical Solutions of Ordinary Differential Equations*

The subject of (ordinary and partial) differential equations is probably the single most important topic in applied mathematics and physical science since most physical processes are described (or modeled) by equations. Unfortunately, these equations do not usually yield exact (or closed form) solutions. For this reason the use of numerical solutions of differential equations is necessary. The introduction of increasingly faster, modern computers greatly increases the importance of numerical methods since more and more problems can be efficiently solved by numerical methods.

There are at least four types of people involved in solving real world problems. The engineer or physical scientist who determines the model and related equations, the mathematician who decides whether this determination is well-posed, the numerical analyst who chooses numerical methods and algorithms to numerically solve the equations and the computer programmer who implements these methods and algorithms on a computer. In practice all four tasks may be done by one person who often ignores the questions of well-posedness and the suitability of the algorithm.

The purpose of this chapter is to introduce the topic of numerical solutions of ordinary differential equations. We will do this by considering the problem of finding the numerical solution of an initial value problem for a first order differential equation:

$$y'(x) = f(x, y), \quad y(a) = y_a \,. \tag{7.1}$$

We realize that most if not all of this material is new to the student so we will proceed slowly and give only the most basic ideas. We encourage the reader to consult more specialized texts when he/she has to deal with real problems.

We have chosen (7.1) for a variety of reasons including:

(1) It is the setting probably most familiar to the reader.

(2) While we usually expect that the independent variable, y, is a real number, most of our ideas and results immediately follow when y is a vector valued function.

(3) Existence, uniqueness and continuity of the solution $y(x)$ as a function of the initial condition y_a are guaranteed by formal theory with little smoothness on f. This is not true for boundary value problems.

(4) Higher order equations, involving derivatives $y''(x)$, $y'''(x)$, ... can be converted to a first order system described in (2).

The reader has already encountered elementary examples of differential equations which yield exact (or closed form) solutions. For example, if $f(x, y)$ is a constant c, then $y'(x) = c$ leads to solutions

$$y(x) = cx + d$$

where d is formed from

$$y_a = y(a) = ca + d \,,$$

that is,

$$d = y_a - ca \,.$$

This gives

$$y(x) = cx + y_a - ca = c(x - a) + y_a \,.$$

Thus, the solution of (7.1) in this case is the line with slope c passing through the point (a, y_a). Similary, the reader has seen the equation of growth or decay

$$y'(x) = ky$$

where

$$f(x, y) = ky$$

in (7.1) and k is a constant. We recall that in this case the solution in (7.1) is

$$y(x) = y_a e^{k(x-a)}.$$

If $k > 0$ our physical problem involves exponential growth such as in bank balances or population increase. If $k < 0$ our physical problem involves exponential decay such as the half-life disintegration of carbon for use in carbon dating or the loss of temperature of an object to its surroundings as in *Newton's law of cooling.* In addition to these examples which are found in most calculus books, the reader may soon encounter problems dealing with the motion of pendulums, spring models, electrical circuits or equations of motion involving higher order derivatives of $y(x)$. They can often be solved by methods discussed in Section 7.5.

Problem 7.1: Show that $y(x) = c(x-a)+y_a$ and $y(x) = y_a e^{k(x-a)}$ each satisfy (7.1) with $f(x, y) = c$ and $f(x, y) = ky$, respectively.

In Section 7.1 we will give some introductory and intuitive ideas for our problem (7.1). In Section 7.2 we will develop the usual single-step methods which are used to solve (7.1). In Section 7.3, we will introduce multi-step methods to solve (7.1). These methods are sometimes troublesome because they are hard to derive and require additional starting values, but they provide higher order algorithms and require fewer new evaluations. In Section 7.4, we will give predictor-corrector methods which allow us to guess the solution and then correct our guess. They also allow us to estimate the errors we have and to choose a more appropriate step size h if necessary. In Section 7.5, we will consider a first order system

$$Y'(x) = F(x, Y), \quad Y(a) = Y_a \tag{7.2}$$

where Y and F are n-vectors. These methods allow us to solve higher order equations, by turning these higher order equations into a system of the form (7.2).

7.1 Introductory Ideas

We begin by trying to make sense of the equation

$$y'(x) = f(x, y) \tag{7.3}$$

and the initial condition

$$y(a) = y_a\,. \tag{7.4}$$

Equation (7.3) is an equation with independent variable x and dependent variable $y = y(x)$. We assume f is a "nice" function of the variables x and y, which yields, when evaluated at the point (x, y), a number which is the slope $y'(x)$ of the solution curve $y(x)$. We ask for a solution on some interval $[a, b)$ where we know the initial value y_a of $y(x)$ at a. Since only one derivative of $y(x)$ is involved in (7.3), we say that we have a first order differential equation.

For example, if

$$y'(x) = xy(x), \quad y(0) = 1 \tag{7.5}$$

we can verify that $y(x) = e^{\frac{1}{2}x^2}$ satisfies (7.5) since

$$y'(x) = \frac{1}{2}(2x)e^{\frac{1}{2}x^2} = xe^{\frac{1}{2}x^2}$$

and $y(0) = 1$. It is instructive to picture what is happening. At each point close to $x = 0$ and $y = 1$ we draw a slope field where the slope

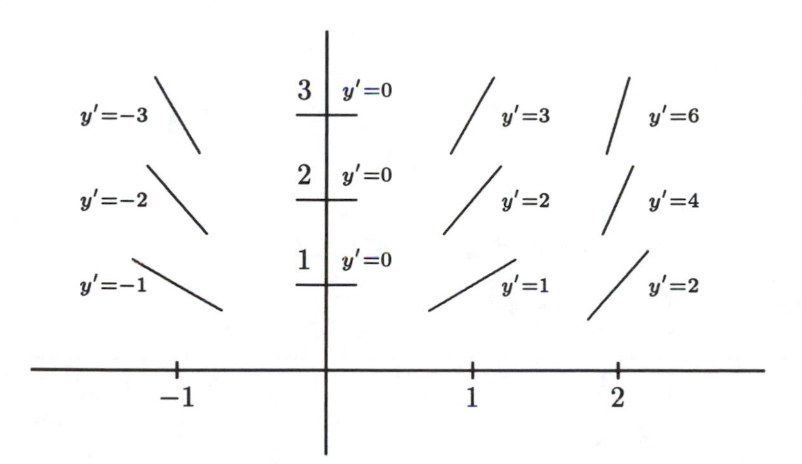

Figure 7.1

at each point (x, y) satisfies $f(x, y) = xy$. This has been done in Figure 7.1, above, for $x = -1, 0, 1, 2$ and $y = 1, 2$, and 3.

Starting at (0,1), which is the initial condition, we attempt to draw a smooth curve for $x > 0$, so that the curve $y(x)$ goes through (0,1) and has a slope $y'(x)$ whose value is given by the slope field at each point (x, y). If we could do this process infinitesimally, in the limit as $h \to 0$, we would obtain the solution $y(x) = e^{\frac{1}{2}x^2}$.

In fact we cannot do this process infinitesimally. In real life we need a rule or method which gives numerical values y_n which approximate the values of the solution $y(x)$ of (7.3), (7.4) at the points $x = x_n$. Table 7.1 in Section 7.2 gives these results for the method called Euler's method. For now we confine ourselves to a picture (see Figure 7.2, below).

In this example $h = \frac{1}{8}$ is the step size and $x_n = x_0 + nh = \frac{n}{8}$ for $n = 0, 1 \ldots, 8$. Euler's method generates the points (x_n, y_n) which approximate the values $(x_n, y(x_n))$ we seek. In our picture the upper curve represents the solution $y(x) = e^{\frac{1}{8}x^2}$ of problem (7.5). The lower curve which is to represent the numerical solution is misleading as we have added (interpolated) smooth curves between successive points

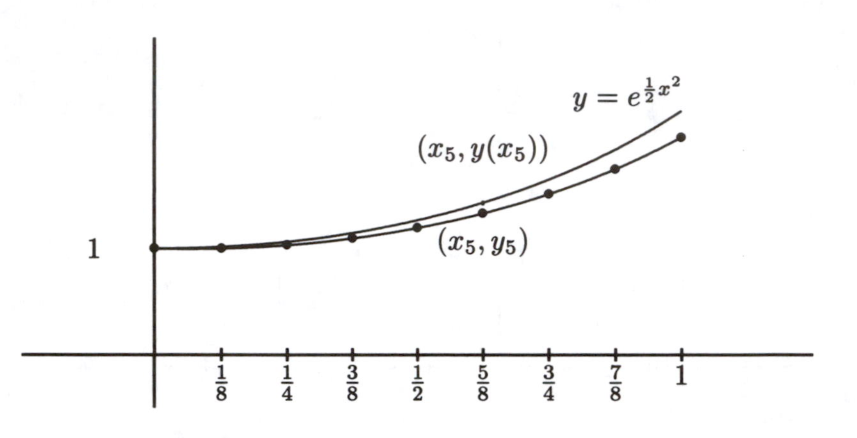

Figure 7.2

(x_n, y_n). In theory, assuming no round-off error, as $h \to 0$ the curves $y_h(x)$ converge to the actual solution $y(x) = e^{\frac{1}{2}x^2}$.

As we have indicated above, we must have some justification to believe that a solution to (7.3), (7.4) exists and is unique. The following theorem appears in any textbook on ordinary differential equations. It is a sufficiency theorem and requires certain smoothness conditions on f.

Theorem 7.1: *If f and $\frac{\partial f}{\partial x}$ are continuous in a rectangle*

$$R = \{|x - a| \leq \delta, |y - y_a| \leq \epsilon : \delta > 0, \epsilon > 0\}$$

then there is an interval $|x - a| \leq h < \delta$ in which there exists a unique solution $y = \phi(x)$ of (7.3), (7,4).

We note that Theorem 7.1 is usually proven by Picard's Method. (See Exercises 7.1 and 7.2.) An immediate corollary of this proof is that the solution depends continuously on the initial data. That is, if we change y_a continuously, the resulting solution $y(x)$ (which depends on y_a) is a continuous function of $y_a(a)$.

7.2 Taylor Series and Runge-Kutta Methods

The purpose of this section is to derive several of the most popular and easiest to apply methods to solve the initial value problem

$$y'(x) = f(x, y(x)), \quad y(a) = y_a \tag{7.6}$$

and the errors that are associated with these methods. These methods are called *one-step methods* because the numerical approximation y_{n+1} of the actual solution $y(x_{n+1})$ is found from the computed value of y_n and the function $f(x, y)$. Throughout this section we will assume that f is smooth enough for our purposes and write $y(x_n)$ for the actual solution value at $x = x_n$ and y_n as the corresponding numerical solution given by our algorithm.

The simplest method is derived from the Taylor series expansion

$$y(x_{n+1}) = y(x_n) + hy'(x_n) + \frac{h^2}{2}y''(\xi_n) \tag{7.7}$$

where $h = x_{n+1} - x_n$ and ξ_n satisfies $x_n < \xi_n < x_n + h$. The algorithm

$$y_{n+1} = y_n + hf(x_n, y_n) \tag{7.8}$$

is called *Euler's Method.* In the next few paragraphs we will see that the local truncation error is $O(h^2)$ for this algorithm and the global error is $O(h)$.

A complete error analysis for the Euler method becomes rather technical, but we can present the major ideas rather easily. We begin by assuming that there is no round-off error in our calculations using (7.8). This assumption is not really true if h is very small, but for the step size we usually use, the round-off error is so small that it can be ignored. Thus, the major source of error is the local truncation error.

The *local truncation error* is the error $\epsilon_{n+1} = y(x_{n+1}) - y_{n+1}$ we get in using (7.8) assuming no other errors. Thus, using (7.7) and (7.8) we have

$$\begin{aligned} \epsilon_{n+1} &= y(x_{n+1}) - y_{n+1} \\ &= y(x_n) + hy'(x_n) + \frac{h^2}{2}y''(\xi_n) - \big(y_n + hf(x_n, y_n)\big) \\ &= \frac{h^2}{2}y''(\xi_n) \end{aligned} \tag{7.9}$$

since by our assumption $y(x_n)$ is the same as y_n and $y'(x_n)$ is the same as $f(x_n, y_n)$. Assuming that $y''(x)$ is bounded on the interval $[a, b]$ we write $\epsilon_{n+1} = O(h^2)$.

The second error is the global or accumulation error. For convenience, we assume a partition of $[a, b]$ into equally spaced points $x_0 = a < x_1 < \cdots < x_N = b$ where $x_{n+1} = x_n + h$. The *global error* is the value of $G = y(b) - y_N$. It can be shown using mathematics beyond the scope of this text that for Euler's method we have $y(b) - y_N = O(h)$. Intuitively, the global error is the sum of the local truncation errors. Thus, using this intuitive definition,

$$\begin{aligned} G &= \sum_{n=0}^{N-1} \epsilon_{n+1} = \sum_{n=0}^{N-1} Ch^2 = NCh^2 \\ &= NhCh = (b-a)Ch = C_1 h \end{aligned} \tag{7.10}$$

and $G = O(h)$. We remind the reader that our error results though intuitively derived are correct. It can be shown that the following uniform error estimate holds:

$$\max_{1 \le n \le N} |y(x_n) - y_n| \le \frac{Mh}{2L}(e^{L(b-a)} - 1) \tag{7.11}$$

where $|y''(x)| \le M$ on $[a, b]$ and L is a Lipschitz constant for f, that is $|f(x, y_1) - f(x, y_2)| \le L|y_1 - y_2|$ for all points (x, y_1) and (x, y_2) in the domain of f.

As an example we consider the problem

$$y'(x) = xy(x), \quad y(0) = 1. \tag{7.12}$$

Table 7.1 contains the numerical calculations for this example. The first column is the value of the independent variable x, the second column is the exact value of the solution $y(x) = e^{\frac{x^2}{2}}$ at the corresponding x values. The third column gives the calculated values using Euler's method with step size $h = \frac{1}{8}$ and the fourth column gives the error between the calculated value and the exact value. The fifth and sixth columns and the seventh and eighth columns give analogous results for step size $\frac{h}{2} = \frac{1}{16}$ and $\frac{h}{4} = \frac{1}{32}$, respectively. The reader should note that the values in the sixth column are approximately one half the respective values in the fourth column and twice the respective values in the eighth column. This illustrates the fact that the error is $O(h)$. Finally, in the last four columns, we give the Richardson values $y_h^R(h) = 2y_{h/2}(x) - y_h(x)$ and the associated respective errors. Note that the error values in column 12 are approximately one-fourth of the respective error values in column 10 since the Richardson error is $O(h^2)$. (See Exercise 7.5.)

Problem 7.2: Derive an expression for the Richardson error $E_k^R(h) = y(x_k) - y_k^R(h)$ in column 10 of Table 7.1. Do most of the results in column 12 of Table 7.1 appear correct?

Higher order Taylor series methods can be found as follows. We write

$$y(x+h) = y(x) + hy'(x) + \frac{h^2}{2}y''(x) + \frac{h^3}{3!}y'''(x) + \cdots \tag{7.13}$$

Euler's Method with Step Size $h = \frac{1}{8}$

x_k	$y(x_k)$	$y_k(h)$	$y(x_k)-y_k(h)$	$y_k\left(\frac{h}{2}\right)$	$y(x_k)-y_k\left(\frac{h}{2}\right)$
0.000	1	1	0	1	0
0.125	1.00784307	1	7.84307E-03	1.00390625	3.93682E-03
0.250	1.03174339	1.015625	0.01611839	1.02360569	8.1377E-03
0.375	1.07284337	1.04736328	0.02548009	1.05990419	0.01293918
0.500	1.13314843	1.09645843	0.03669	1.1144067	0.01874173
0.625	1.21569081	1.16498708	0.05070373	1.18963458	0.02605623
0.750	1.32478474	1.25600169	0.06878305	1.28921855	0.03556619
0.875	1.46640305	1.37375184	0.09265121	1.4181876	0.04821545
1.000	1.64872125	1.52400594	0.12471531	1.58338602	0.06533523

$y_k\left(\frac{h}{4}\right)$	$y(x_k)-y_k\left(\frac{h}{4}\right)$	$y_k^R(h)$	$y(x_k)-y_k^R(h)$	$y_k^R\left(\frac{h}{2}\right)$	$y(x_k)-y_k^R\left(\frac{h}{2}\right)$
1	0	1	0	1	0
1.00586986	1.97321E-03	1.0078125	3.057E-05	1.00783347	9.6E-03
1.02765263	4.09076E-03	1.03158638	1.5701E-04	1.03169957	4.382E-05
1.06631964	6.52373E-03	1.0724451	3.9827E-04	1.07273509	1.0828E-04
1.12367053	9.4779E-03	1.13235497	7.9346E-04	1.13293436	2.1407E-04
1.20247278	0.01231803	1.21428208	1.40873E-03	1.21531098	3.7983E-04
1.306684	0.1810074	1.32243541	2.34933E-03	1.32414945	6.3529E-04
1.44178176	0.02462129	1.46262336	3.77969E-03	1.46537592	1.02713E-03
1.61523932	0.03348193	1.6427661	5.95515E-03	1.64709262	1.62863E-03

Note: $y_k\left(\frac{h}{4}\right)$ is the seventh column of Table 7.1.

Table 7.1

and note that the following hideous formal calculations can be made by using the chain rule

$$\begin{aligned}
y'(x) &= f(x, y(x)),\\
y''(x) &= \frac{d}{dx} f(x, y(x))\\
&= f_x + f_y y'(x)\\
&= f_x + f_y f,\\
y'''(x) &= \frac{\partial}{\partial x}(f_x + f_y f) + \left(\frac{\partial}{\partial y}(f_x + f_y f)\right) y'\\
&= f_{xx} + f_{xy} f + f_y f_x + (f_{xy} + f_{yy} f + f_y^2) y'\\
&= f_{xx} + 2 f_{xy} f + f_y f_x + f_{yy} f^2 + f_y^2 f,\\
&\vdots
\end{aligned}$$

where the functions f, $f_{x'}$, $f_{y'}$, $f_{xy} = f_{yx}$ etc. are evaluated at the point $(x, y(x))$. For example, using the first two equations above we have the following algorithm

$$\begin{aligned}
y_{n+1} &= y_n + hf(x_n, y_n)\\
&\quad + \frac{h^2}{2}[f_x(x_n, y_n) + f_y(x_n, y_n) f(x_n, y_n)]\,.
\end{aligned} \tag{7.14}$$

Similar reasoning to Euler's method gives a local truncation error for ϵ_{n+1} in (7.14) of $O(h^3)$ and a global error of $O(h^2)$.

For our example in (7.12) where $y' = xy$, we have

$$y'' = (xy)' = y + xy' = y + x^2 y$$

so that (7.14) becomes

$$y_{n+1} = y_n + h[x_n y_n] + \frac{h^2}{2}(y_n + x_n^2 y_n)\,. \tag{7.15}$$

Our computer results in Table 7.2, below, are presented in the same form as Table 7.1. Note that in this table the error values in column 6 are approximately one-fourth of the respective values in column 4. Although we have not given the Richardson values for $\frac{h}{2}$, the Richardson error values would be of order $O(h^3)$. (See Exercise 7.6.)

Problem 7.3: Repeat the example problem of the last section to obtain a Taylor series algorithm with local truncation error $O(h^4)$.

Problem 7.4: Give a Taylor series algorithm similar to (7.14) which has a local truncation error of $O(h^4)$.

Our next topic is Runge-Kutta methods. The reader can see that to achieve a higher order of accuracy for Taylor series methods we will have to find various derivatives $y''(x)$, $y'''(x)$, ... This is an unpleasant and often impossible task. Fortunately, we may replace the derivatives by evaluations of $f(x,y)$ at intermediate points so as to achieve the same desired accuracy. Methods derived in this way are called *Runge-Kutta methods.* Our first task is to derive a Runge-Kutta method of order two (local truncation error of order three) to replace the Taylor series method in (7.14). Then we will give the most common Runge-Kutta method of order four and invite the reader to do the necessary calculations.

The basic idea is to replace the approximation

$$\begin{aligned} y(x+h) &= y(x) + hy'(x) + \frac{h^2}{2}y''(x) + O(h^3) \\ &= y(x) + hf(x,y) \\ &\quad + \frac{h^2}{2}[f_x(x,y) + f_y(x,y)f(x,y)] + O(h^3) \end{aligned}$$

by

$$y(x+h) = y(x) + w_1k_1 + w_2k_2$$

where $k_1 = hf(x,y)$ and $k_2 = hf(x+\alpha h, y+\beta k_1)$ and w_1, w_2, α, and β are chosen to preserve the order of approximation. The motivation is that k_1 and k_2 are aproximations to Δy. Thus $y_n + k_1$

Second Order Taylor Series with $h = \frac{1}{8}$

x_k	$y(x_k)$	$y_k(h)$	$y(x_k)-y_k(h)$	$y_k\left(\frac{h}{2}\right)$	$y(x_k)-y_k\left(\frac{h}{2}\right)$	$y_k^R(h)$	$y(x_k)-y_k^R(h)$
0.000	1	1	0	1	0	1	0
0.125	1.00784307	1.0078125	3.057E-05	1.00783157	1.15E-03	1.007837927	5.143E-06
0.250	1.03174339	1.03155612	1.8727E-04	1.03168822	5.517E-05	1.031732253	1.1137E-05
0.375	1.07284337	1.07235496	4.8841E-04	1.07270705	1.3632E-04	1.072824413	1.8957E-05
0.500	1.13314843	1.13217748	9.7095E-04	1.13288322	2.6521E-04	1.133118467	2.9963E-05
0.625	1.21569081	1.21399499	1.69582E-03	1.21523231	4.585E-04	1.21564475	4.606E-05
0.750	1.32478474	1.32202749	2.75725E-03	1.3240429	7.4814E-04	1.324714703	7.0037E-05
0.875	1.46640305	1.46210559	4.29746E-03	1.46524905	1.154E-03	1.46629687	1.0618E-04
1.000	1.64872125	1.64219158	6.52967E-03	1.64696808	1.75317E-03	1.648560247	1.61003E-04

Note: $y_k^R(h)$ is the seventh column of Table 7.2.

Table 7.2

or $y_n + \frac{1}{2}k_1 + \frac{1}{2}k_2$, as in (7.16), are better approximations to $y(x_{n+1})$ then the approximation used in (7.8). Expanding k_2 about the point (x, y) we have

$$k_2 = h\left(f(x,y) + \alpha h f_x(x,y) + \beta k_1 f_y(x,y) + O(h^2)\right)$$

so that

$$\begin{aligned} y(x+h) &= y(x) + w_1 h f(x,y) \\ &\quad + w_2 h\big[f(x,y) + \alpha h f_x(x,y) + \beta h f(x,y) f_y(x,y)\big] + O(h^3)\,. \end{aligned}$$

Equating our two expressions with error $O(h^3)$ we have

$$w_1 + w_2 = 1\,, \quad w_2\alpha = \frac{1}{2} \quad \text{and} \quad w_2\beta = \frac{1}{2}\,.$$

This gives three equations in the four unknowns w_1, w_2, α and β which gives one degree of freedom so that there are an infinite number of solutions to our problem. However, it is usually best to take the weights w_1 and w_2 as equal as possible. Thus, we choose $w_1 = w_2 = \frac{1}{2}$ and $\alpha = \beta = 1$ and obtain the following theorem.

Theorem 7.2: *The Runge-Kutta algorithm*

$$\begin{aligned} y_{n+1} &= y_n + \tfrac{1}{2}k_1 + \tfrac{1}{2}k_2 \\ k_1 = hf(x_n, y_n)\,, &\quad k_2 = hf(x_{n+1}, y + k_1) \end{aligned} \tag{7.16}$$

has a local truncation error equal to $O(h^3)$ *which is the same local truncation error as given by the Taylor series method* (7.14).

The best known Runge-Kutta method is given below in (7.12) and has local truncation error equal to $O(h^5)$. It is

$$(7.17)\qquad \begin{cases} y_{n+1} = y_n + \frac{1}{6}(k_1 + 2k_2 + 2k_3 + k_4) \text{ where} \\ k_1 = hf(x_n, y_n), \\ k_2 = hf\left(x_n + \frac{h}{2}, y_n + \frac{k_1}{2}\right), \\ k_3 = hf\left(x_n + \frac{h}{2}, y_n + \frac{k_2}{2}\right) \text{ and} \\ k_4 = hf(x_{n+1}, y_n + k_3). \end{cases}$$

The reader may derive (7.17) as we have derived (7.16). In this case, we have weights $w_1 = \frac{1}{6}$, $w_2 = \frac{2}{6}$, $w_3 = \frac{2}{6}$, and $w_4 = \frac{1}{6}$ and $\alpha_1 = \alpha_2 = \frac{1}{2}$, $\alpha_3 = 1$, $\beta_1 = \beta_2 = \frac{1}{2}$ and $\beta_3 = 1$. If this is too much work, the reader may show that (7.17) reduces to the Taylor series algorithm with error $O(h^5)$.

Problem 7.5: Justify the last sentence above.

Problem 7.6: Derive (7.17).

Finally, we note that if $f(x, y) = g(x)$, that is, $y'(x) = g(x)$ where g is independent of y, then the approximation to

$$y(x_{n+1}) - y(x_n) = \int_{x_n}^{x_{n+1}} y'(x)\, dx$$

given by (7.17) is

$$y_{n+1} = y_n + \frac{1}{6}\left[hg(x_n) + 2hg(x_n + \frac{h}{2}) + 2hg(x_n + \frac{h}{2}) + hg(x_n + h)\right]$$

which is Simpson's rule. In a similar manner, if $f(x, y)$ is independent of y, these one step methods may be used to obtain the quadrature rules.

Table 7.3 contains the results for our standard example $y' = xy$, $y(0) = 1$ using the fourth order Runge-Kutta algorithm. We expect that the errors in the sixth column would be $\frac{1}{16}$ as large as those in the fourth column. This would be the case (for example) if our calculations were done on a machine which carries more than eight decimal places. Hence we are observing the effect of round-off error.

Fourth Order Runge-Kutta with $h = \frac{1}{8}$

x_k	$y(x_k)$	$y_k(h)$	$y(x_k)-y_k(h)$	$y_k\left(\frac{h}{2}\right)$	$y(x_k)-y_k\left(\frac{h}{2}\right)$
0.000	1	1	0	1	0
0.125	1.00784307	1.00784309	-2E-08	1.00784309	-2E-08
0.250	1.03174339	1.03174339	0	1.03174339	0
0.375	1.07284337	1.07284336	1E-08	1.07284337	0
0.500	1.13314843	1.1331484	3E-08	1.13314842	1E-08
0.625	1.21569081	1.21459074	7E-08	1.21569078	3E-08
0.750	1.32478474	1.32478458	1.6E-07	1.32478469	5E-08
0.875	1.46640305	1.46640269	3.6E-07	1.46640296	9E-08
1.000	1.64872125	1.64872053	7.2E-07	1.64872113	1.2E-07

Table 7.3

7.3 Multistep Methods

The purpose of this section is to consider methods of the form

$$\begin{aligned} y_{n+1} &= \alpha_1 y_n + \alpha_2 y_{n-1} + \cdots + \alpha_k y_{n-k+1} \\ &\quad + h\big[\beta_0 f(x_{n+1}, y_{n+1}) + \beta_1 f(x_n, y_n) + \cdots \\ &\quad + \beta_k f(x_{n-k+1}, y_{n-k+1})\big] \end{aligned} \tag{7.18}$$

to solve our first order, initial value problem,

$$y'(x) = f(x, y(x)), \quad y(a) = y_a \,. \tag{7.19}$$

Equation (7.18) is called a *multistep method* for (7.19). More precisely, we say that (7.18) is a k-step method. If $\beta_0 \neq 0$, the method is said to be *implicit*. In this case, y_{n+1} appears on both sides of (7.1) and we will have to solve for y_{n+1} using one of the methods from Chapter 3. If $\beta_0 = 0$ the method is called *explicit*. We note that the sum of the α's and the β's must be equal to one. (See Exercise 7.27.)

The major advantage to multistep methods is that we usually obtain higher accuracy with less work than for single step methods. There are several important disadvantages however. One is that multistep methods are not self starting as are single step methods, since in order to apply (7.18), we require $y_n, y_{n-1}, \ldots, y_{n-k+1}$ for a k-step method.

For example, for an explicit 2-step method, we need the values $y_0 = y_a$ and y_1 to make our first computation using (7.18), which is,

$$y_2 = \alpha_1 y_1 + \alpha_2 y_0 + h\big[\beta_1 f(x_1, y_1) + \beta_2 f(x_0, y_0)\big] \,.$$

The problem is that while we may have y_0 from (7.19), some other method will be required to find y_1. Care must be taken that these starting values are accurate enough for the method we use.

Multistep methods are usually derived using interpolating polynomials in a manner similar to Newton-Cotes methods. To see how a two-step explicit method is derived we start with the Lagrange interpolating polynomial at the points (x_{n-1}, y_{n-1}) and (x_n, y_n) for the function $g(x) = f(x, y(x))$ where $y(x)$ is the solution to (7.19) and y_{n-1} and y_n are the last two values found by (7.18). Thus,

$$\begin{aligned} g(x) = {} & \frac{x - x_n}{x_{n-1} - x_n} g(x_{n-1}) + \frac{x - x_{n-1}}{x_n - x_{n-1}} g(x_n) \\ & + \frac{1}{2}(x - x_{n-1})(x - x_n) g''(\xi_n) \end{aligned} \tag{7.20}$$

where $x_{n-1} < \xi_n < x_{n+1}$.

Using the Fundamental Theorem of Calculus, (7.19) and (7.20), we have

$$\begin{aligned}
y(x_{n+1}) - y(x_n) &= \int_{x_n}^{x_{n+1}} y'(x)\,dx \\
&= \int_{x_n}^{x_{n+1}} f(x, y(x))\,dx \\
&= -\frac{g(x_{n-1})}{h} \int_{x_n}^{x_{n+1}} (x - x_n)\,dx \\
&\quad + \frac{g(x_n)}{h} \int_{x_n}^{x_{n+1}} (x - x_{n-1})\,dx \\
&\quad + \frac{1}{2} \int_{x_n}^{x_{n+1}} (x - x_{n-1})(x - x_n) y''(\xi_x)\,dx \\
&= -\frac{f(x_{n-1}, y_{n-1})}{h} \left(\frac{h^2}{2}\right) + \frac{f(x_n, y_n)}{h} \left(\frac{3h^2}{2}\right) + E \\
&= h\left(-\frac{1}{2} f(x_{n-1}, y_{n-1}) + \frac{3}{2} f(x_n, y_n)\right) + E
\end{aligned}$$

where

$$\begin{aligned}
E &= \frac{1}{2} \int_{x_n}^{x_{n+1}} (x - x_{n-1})(x - x_n) g''(\xi_x)\,dx \\
&= \frac{1}{2} g''(\xi_n) \int_{x_n}^{x_{n+1}} (x - x_{n-1})(x - x_n)\,dx \\
&= \frac{1}{2} g''(\xi_n) \int_{h}^{2h} s(s - h)\,ds \\
&= \frac{5}{12} h^3 g''(\xi_n) = \frac{5}{12} h^3 y'''(\xi_n)
\end{aligned}$$

is the local error and $x_n < \xi_n < x_{n+1}$. To derive E, we have used the Integral Means theorem, Theorem 2.8 of Section 2.2, since

$$h(x) = (x - x_{n-1})(x - x_n) \geq 0$$

on $[x_n, x_{n+1}]$.

The explicit, two-step algorithm we have just derived is called the *Adams-Bashforth two-step method*:

$$y_{n+1} = y_n + h\left(\frac{3}{2}f(t_n, y_n) - \frac{1}{2}f(t_{n-1}, y_{n-1})\right) \tag{7.21}$$

with associated global error

$$G = \frac{5}{12}h^2 y'''(\xi), \tag{7.22}$$

where $a < \xi < b$. Comparison with (7.18) yields $\alpha_1 = 1$, $\alpha_2 = 0$, $f_0 = 0$, $\beta_1 = \frac{3}{2}$, and $\beta_2 = -\frac{1}{2}$. As before, the global error can be intuitively found by summing the local errors. Hence

$$G = \sum_{n=0}^{N-1} Ch^3 = NCh^3 = C(b-a)h^2 .$$

We note that the two-step implicit method would begin with the expression

$$\begin{aligned} g(x) \approx & \frac{(x-x_n)(x-x_{n+1})}{(x_{n-1}-x_n)(x_{n-1}-x_{n+1})} g(x_{n-1}) \\ & + \frac{(x-x_{n-1})(x-x_{n+1})}{(x_n-x_{n-1})(x_n-x_{n+1})} g(x_n) \\ & + \frac{(x-x_{n-1})(x-x_n)}{(x_{n+1}-x_{n-1})(x_{n+1}-x_n)} g(x_{n+1}) \\ & + \frac{1}{6}(x-x_{n-1})(x-x_n)(x-x_{n+1}) g'''(\xi_x) \end{aligned}$$

so that the local error corresponding to (7.22) would be $O(h^4)$. Thus, we pick up one order of accuracy by using an implicit method. In fact, if we carried out these calculations we would obtain the *Adams-Moulton two-step method*

$$y_{n+1} = y_n + f\left(\frac{5}{12}f(x_{n+1}, y_{n+1}) + \frac{8}{12}f(x_n, y_n) - \frac{1}{12}f(x_{n-1}, y_{n-1})\right) \tag{7.23}$$

with global error

$$E = -\frac{1}{24}h^3 y^{(iv)}(\xi)\,. \tag{7.24}$$

Problem 7.7: Derive (7.23) and (7.24).

These results illustrate the general case. The (explicit) Adams-Bashforth m-step method gives the same order of global error, $O(h^m)$, as the (implicit) Adams-Moulton $m-1$ step method. Each method requires several evaluations of f, of which all but one have been done before. The problem with implicit methods is that we must solve a nonlinear equations. In theory these nonlinear equations can be solved by methods such as Newton's method. In practice we usually prefer to use explicit methods to avoid the extra computations. The implicit methods enjoy the advantage of greater stability as the respective β coefficients of the implicit methods are generally smaller then the β coefficients of the explicit methods. We note that the Adams-Bashforth three-step method is

$$y_{n+1} = y_n + h\left(\frac{23}{12}f(x_n, y_n) - \frac{16}{12}f(x_{n-1}, y_{n-1}) + \frac{5}{12}f(x_{n-2}, y_{n-2})\right) \tag{7.25}$$

with global error

$$E = \frac{3}{8}h^3 y^{(iv)}(\xi)\,. \tag{7.26}$$

The reader may appreciate that an infinite number of multistep methods can be derived. In the next section we will consider the most popular ones which are used in predictor-corrector methods.

Finally, in Table 7.4, we present numerical calculations for the solution of our standard example

$$y'(x) = xy, \quad y(0) = 1\,,$$

based on the Adams-Bashforth two-step method ((7.21) above). As usual, the first column presents the x values and the second column presents the corresponding exact solution. Since the algorithm is a two-step method we must use some one-step method to calculate $y_1(h)$. The next six columns present the calculations based on three different single step methods. Columns 3 and 4 give the calculated values and the corresponding error if we use Euler's method (7.8) of Section 7.2 to calculate $y_1(h)$. Columns 5 and 6 give the calculated values and the corresponding error if we use the second order Taylor series method ((7.14) of Section 7.2) to calculate $y_1(h)$. Since the Adams-Bashforth two-step method is of order two, as is the Taylor series method, the errors given in column 6 are much better than the errors given in column 4 where Euler's method is used to calculate the initial value. Finally, columns 7 and 8 give the calculated values and the corresponding error if we use the fourth order Runge-Kutta method (7.17 of Section 7.2) to calculate $y_1(h)$. Note that even though Runge-Kutta is fourth order, the betterment in error over using the second order Taylor series is negligible.

Adams-Bashforth Second Order
with Different Starters and $h = 0.2$

Euler's Method

x_k	$y(x_k)$	$y_k(h)$	$y(x_k)-y_k(h)$
0.0	1	1	0
0.2	1.02020132	1	0.02020132
0.4	1.08328705	1.06	0.02328705
0.6	1.08328705	1.1672	0.03001735
0.8	1.37712776	1.334896	0.04223176
1.0	1.64872125	1.58523904	0.06348221

Second Order Taylor Series		Fourth Order Runge-Kutta	
$y_k(h)$	$y(x_k)-y_k(h)$	$y_k(h)$	$y(x_k)-y_k(h)$
1	0	1	0
1.0202	1.33E-05	1.02020133	-1E-08
1.0812	2.08705E-03	1.0814134	1.87365E-03
1.190544	6.67335E-03	1.19077898	6.43847E-03
1.36159392	0.01553384	1.36186266	0.0152651
1.61694381	0.03177744	1.61726295	0.0314583

Table 7.4

7.4 Predictor-Corrector Methods

In this section we consider predictor-corrector methods to solve our first order initial equation

$$y'(x) = f(x, y(x)), \quad y(a) = y_a. \tag{7.27}$$

These are often the methods which are used for large canned subroutine packages as found in industrial companies. Their major advantage is that they allow us to "check" our errors and hence to change or adapt the step size h to achieve a specified degree of accuracy as we move from point to point for a variety of functions $f(x, y)$.

Generally there are three parts of a predictor-corrector package with a fixed order of error. The first part is a single step method to generate the starting values $y_1, y_2, \ldots, y_{k-1}$ for the second part which is an implicit, multistep, predictor method of the form

$$\begin{aligned} y_{n+1}^{(P)} &= \alpha_1 y_n + \alpha_2 y_{n-1} + \cdots + \alpha_k y_{n-k+1} \\ &\quad + h\big[\beta_1 f(x_n, y_n) + \cdots + \beta_k f(x_{n-k+1}, y_{n-k+1})\big]. \end{aligned} \tag{7.28}$$

This generates a predicted value for y_{n+1} which we denote by $y_{n+1}^{(P)}$ for $n \geq k$. The third part is an implicit, multistep corrector method of the form

$$\begin{aligned} y_{n+1}^{(C)} &= \gamma_1 y_n + \gamma_2 y_{n-1} + \cdots + \gamma_{k-1} y_{n-k+2} \\ &\quad + h\big[\delta_0 f(x_{n+1}, y_{n+1}^{(P)} \\ &\qquad + \cdots + \delta_{k-1} f(x_{n-k+2}, y_{n-k+2})\big]. \end{aligned} \tag{7.29}$$

We call $y_{n+1}^{(C)}$ the corrected value for y_{n+1}.

We note that the three parts or algorithms should have the same order of error $O(h^m)$ so that if (7.28) is a k-step method then (7.29) is a $k-1$ step method since, as we have seen above, implicit methods generally have one additional order of accuracy more than explicit methods. The value $y_0 = y_a$ is the initial value. If (7.28) is a k-step method, then the values $y_1, y_2, \ldots, y_{k-1}$ are obtained from the first part of our package, the one-step algorithm. The first predicted value we compute using (7.28) is $y_k^{(P)}$ which occurs when $n - k + 1 = 0$ or $n = k - 1$. The values $y_1, y_2, \ldots, y_k^{(P)}$ are used in (7.29) to obtain $y_k^{(C)}$. If $|y_k^{(C)} - y_k^{(P)}|$ is sufficiently small we set $y_k = y_k^{(C)}$ and attempt to find y_{k+1}. If this difference is not sufficiently small we begin again with a new, smaller value of h (say $\frac{h}{2}$) and repeat the whole process to this point.

Problem 7.8: Justify the first sentence of the last paragraph. Specifically, if $y_{n+1}^{(P)}$ and $y_{n+1}^{(C)}$ are found by algorithms with error $O(h^m)$ explain why the initial values should not be found by an algorithm with error $O(h^q)$ where $q < m$ or $q > m$.

If y_k is acceptable we compute $y_{k+1}^{(P)}$, from (7.28) using the values $y_1, y_2, \ldots, y_k$ and $y_{k+1}^{(C)}$ from (7.29) using the values

$$y_2, y_3, \ldots, y_k, y_{k+1}^{(P)}.$$

If $|y_{k+1}^{(C)} - y_{k+1}^{(P)}|$ is sufficiently small we set $y_k = y_k^{(C)}$. Otherwise, we repeat the steps above with a smaller value of k (say $\frac{h}{2}$) until the difference $|y_{k+1}^{(C)} - y_{k+1}^{(P)}|$ is sufficiently small.

The procedure in the last paragraph is repeated for $n = k+1, k+2, \ldots$. In some packages, if $|y_{k+1}^{(C)} - y_{k+1}^{(P)}|$ is too small, the value of h is increased (say $2h$), before an attempt is made to find y_{n+2}. The reader should realize that the value of h depends to a great degree on the local smoothness of the solution $y(x)$. Thus, if $y = y(x)$, and if the solution to (7.27) is as pictured in Figure 7.3, below, then we require much smaller values of h in a neighborhood of c_2 than is required in a neighborhood of c_1 or c_3. If these small

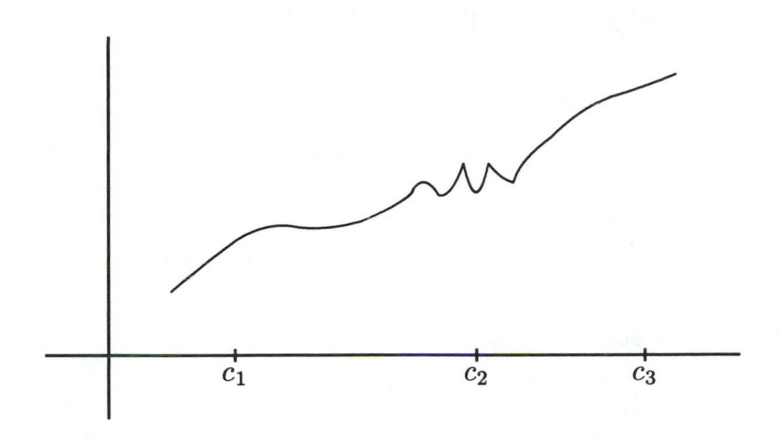

Figure 7.3

values of h are not used in a neighborhood of c_2 it is clear that dire consequence will carry over past $x = c_2$.

Probably, the best example of a predictory-corrector package is to combine the fourth order Runge-Kutta method

$$
(7.30) \qquad \begin{cases} k_1 = hf(t_n, y_n)\,, \\[2ex] k_2 = hf\left(t_n + \dfrac{h}{2}, y_n + \dfrac{k_1}{2}\right)\,, \\[2ex] k_3 = hf\left(t_n + \dfrac{h}{2}, y_n + \dfrac{k_2}{2}\right)\,, \\[2ex] k_4 = hf(t_n + h, y_n + k_3) \quad \text{and} \\[2ex] y_{n+1} = y_n + \dfrac{k_1 + 2k_2 + 2k_3 + k_4}{6} \end{cases}
$$

with the Adams-Bashforth four-step method

$$
\begin{aligned}
y_{n+1}^{(P)} = y_n + \frac{h}{24}\big(&55f(x_n, y_n) - 59f(x_{n-1}, y_{n-1}) \\
&+ 37f(x_{n-2}, y_{n-2}) - 9f(x_{n-3}, y_{n-3})\big)
\end{aligned}
\tag{7.31}
$$

and the Adams-Moulton three-step method

$$
\begin{aligned}
y_{n+1}^{(C)} = y_n + \frac{h}{24}\big[&9f(x_{n+1}, y_{n+1}^{(P)}) + 19f(x_n, y_n) \\
&- 5f(x_{n-1}, y_{n-1}) + f(x_{n-2}, y_{n-2})\big] .
\end{aligned}
\tag{7.32}
$$

All three algorithms have a global error of $O(h^4)$. The value $y_0 = y_a$ is given in (7.27), the values of y_1, y_2, and y_3 are obtained from (7.30). The first value y_{n+1} computed by (7.31) and (7.32) is y_4 when $n = 3$.

In Table 7.5 we present a test run for our standard problem $y' = xy$, $y(0) = 1$. The first column gives the x values with a step size of $h = 0.02$, the second column gives the corresponding exact values $y(x)$, the third and fifth columns give the corresponding calculated values for the fourth order predictor-corrector method and the fourth order Runge-Kutta method respectively. The fourth and sixth columns give the corresponding errors for each method.

Adams Fourth Order Predictor-Corrector vs Runge-Kutta

x_k	$y(x_k)$	y_k^A	$y(x_k)-y_k^A$	y_k^{RK}	$y(x_k)-y_k^{RK}$
0.00	0.9999999998	1.0	-1E-08	1.0	-1E-08
0.02	1.00020001	1.00020002	-1E-08	1.00020002	-1E-08
0.04	1.00080030	1.00080032	-2E-08	1.00080032	-2E-08
0.06	0.00180161	1.00180162	-1E-08	1.00180162	-1E-08
0.08	1.00320512	1.00320512	0	1.00320512	0
0.10	1.00501250	1.00501251	-1E-08	1.00501251	-1E-08
0.12	1.00722596	1.00722597	-1E-08	1.00722597	-1E-08
0.14	1.00984816	1.00984816	0	1.00984816	0
0.16	1.01288226	1.01288225	1E-08	1.01288225	1E-08
0.18	1.01633192	1.01633191	1E-08	1.01633191	1E-08
0.20	1.02020132	1.02020131	1E-08	1.02020131	1E-08
0.22	1.02449519	1.02449516	3E-08	1.02449516	3E-08
0.24	1.02921871	1.02921869	2E-08	1.02921869	2E-08
0.26	1.03437770	1.03437767	3E-08	1.03437767	3E-08
0.28	1.03997845	1.03997841	4E-08	1.03997841	4E-08
0.30	1.04602785	1.04602781	4E-08	1.04602781	4E-08
0.32	1.05253337	1.05253332	5E-08	1.05253332	5E-08
0.34	1.05950305	1.05950301	4E-08	1.05950301	4E-08
0.36	1.06694560	1.06694555	5E-08	1.06694554	6E-08
0.38	1.07487028	1.07487023	5E-08	1.07487022	6E-08
0.40	1.08328705	1.08328700	5E-08	1.08328699	6E-08
⋮					
0.90	1.49930248	1.49930232	1.6E-07	1.49930224	2.4E-07
0.92	1.52683962	1.52683945	1.7E-07	1.52683936	2.6E-07
0.94	1.55550459	1.55550442	1.7E-07	1.55550432	2.7E-07
0.96	1.58534172	1.58534256	1.6E-07	1.58534145	2.7E-07
0.98	1.61639764	1.61639745	1.9E-07	1.61639734	3.0E-07
1.00	1.64872125	1.64872107	1.8E-07	1.64872095	3.0E-07

Table 7.5

7.5 First Order Systems and Higher Order Equations

In this section we consider the solution of a first order system of ordinary differential equations of the form

$$Y'(x) = F(x, Y), \quad Y(a) = Y_a\,, \tag{7.33}$$

where Y and F are m-vectors. We will see that the algorithm and their errors are "exactly" like the case when $m = 1$ in the previous sections. In fact, the only problem for the reader might be in understanding the vector notation involved in (7.33).

As before, we are concerned with an independent variable x defined on the interval $[a, b]$. Instead of one dependent variable $y(x)$ we now have m dependent variable functions $y_1(x), y_2(x), \ldots, y_m(x)$, which are the components of the vector $Y(x)$. For each x in $[a, b]$ we have an m-vector

$$Y(x) = \begin{bmatrix} y_1(x) \\ y_2(x) \\ \vdots \\ y_m(x) \end{bmatrix}, \tag{7.34a}$$

that is, a point in m dimensional space. For ease of presentation we often denote this vector by $[y_1(x), y_2(x), \ldots, y_m(x)]^T$ where T denotes the transpose of the vector. Likewise, for each (x, Y) we assume that there are m real valued functions $f_1(x, Y), f_2(x, Y), \ldots, f_m(x, Y)$. Thus, we write

$$F(x, Y) = \begin{bmatrix} f_1(x, Y) \\ f_2(x, Y) \\ \vdots \\ f_m(x, Y) \end{bmatrix} \tag{7.34b}$$

or $F^T(x, Y)$ or $[f_1(x, Y), f_2(x, Y), \ldots, f_m(x, Y)]^T$ to denote the vector of m real valued functions.

As an example of a first order, ordinary differential equation initial value problem we have

$$(7.35)\quad \begin{gathered} Y'(x) = \begin{bmatrix} y_1'(x) \\ y_2'(x) \\ y_3'(x) \end{bmatrix} = F(x,Y) = \begin{bmatrix} \left(y_2^2(x) + y_3^2(x)\right) e^x \\ y_3(x) \\ -e^{-x} y_1(x) y_2(x) \end{bmatrix}, \\ Y(\pi) = \begin{bmatrix} y_1(\pi) \\ y_2(\pi) \\ y_3(\pi) \end{bmatrix} = \begin{bmatrix} e^\pi \\ 0 \\ -1 \end{bmatrix}. \end{gathered}$$

The reader should verify that $y(x) = [e^x, \sin x, \cos x]^T$ is a solution to this problem near $x = \pi$. There is a general existence and a uniqueness theorem in this case which yields the fact that this is the only solution in the interval $[a, b]$ with $a = \pi$ and $b > \pi$.

Problem 7.9: Verify that $y(x) = [e^x, \sin x, \cos x]^T$ satisfies (7.35).

Problem 7.10: Solve (7.35) by Euler's method given in (7.8) of Section 7.2 for $b = 2\pi$ and $h = \frac{\pi}{7}$.

Problem 7.11: Repeat Problem 7.10 with $h = \frac{\pi}{14}$. Compare your errors with Problem 7.10 and with a Richardson extrapolation using these results.

The methods and the error analysis to solve systems of first order differential equations are analogous to the first order cases presented earlier. For example, we recall the fourth order Runge-Kutta method for $m = 1$, given by

$$\begin{cases} k_1 = hf(x_n, y_n), \\ k_2 = hf\left(x_n + \frac{h}{2},\ y_n + \frac{k_1}{2}\right), \\ k_3 = hf\left(x_n + \frac{h}{2},\ y_n + \frac{k_2}{2}\right), \\ k_4 = hf(x_n + h, y_n + k_3) \quad \text{and} \\ y_{n+1} = y_n + \dfrac{k_1 + 2k_2 + 2k_3 + k_4}{6}. \end{cases}$$

In the case of a system of m equations we get the following adaptation:

$$\begin{aligned}
k_{1i} &= hf_i(x_j, y_{1j}, y_{2j}, \dots, y_{mj}) \\
k_{2i} &= hf_i(x_j + \frac{h}{2}, y_{ij} + \frac{k_{11}}{2}, y_{2j} + \frac{k_{21}}{2}, \dots y_{mj} + \frac{k_{1m}}{2} \\
k_{3i} &= hf_i(x_j + \frac{h}{2}, y_{ij} + \frac{k_{21}}{2}, y_{2j} + \frac{k_{22}}{2}, \dots y_{mj} + \frac{k_{2m}}{2} \qquad \text{and} \\
k_{4i} &= hf_i(x_j + h, y_{ij} + k_{31}, y_{2j} + k_{32}, \dots, y_{mj} + k_{3m}),
\end{aligned}$$

for each $i = 1, 2, \dots, m$, with

$$y_{i,j+1} = y_{ij} + \frac{k_{1i} + 2k_{2i} + 2k_{3i} + k_{4i}}{6}$$

for each $i = 1, 2, \dots, m$. It should be noted that before we calculate k_{21}, for example, we must calculate all of the $k_{11}, k_{12}, \dots, k_{1m}$. If we use our example above with $m = 3$ we have

$$\begin{aligned}
k_{11} &= hf_1(x_j, y_{ij}, y_{2j}, y_{3j}) \\
&= h\left(y_{2j}^2 + y_{3j}^2\right)e^{x_j} \\
k_{12} &= hf_2(x_j, y_{1j}, y_{2j}, y_{3j}) \\
&= hy_{3j} \qquad \text{and} \\
k_{13} &= hf_3(x_j, y_{1j}, y_{2j}, y_{3j}) \\
&= -he^{-x_j}y_{1j}y_{2j}
\end{aligned}$$

with similar expressions for k_{21}, k_{22}, k_{23}, k_{31}, k_{32}, k_{33}, k_{41}, k_{42}, and k_{43}. Then, we would get

$$\begin{aligned}
y_{1,j+1} &= y_{1j} + \frac{k_{11} + 2k_{21} + 2k_{31} + k_{41}}{6} \\
y_{2,j+1} &= y_{2j} + \frac{k_{12} + 2k_{22} + +2k_{32} + k_{42}}{6} \qquad \text{and} \\
y_{3,j+1} &= y_{3j} + \frac{k_{13} + 2k_{23} + 2k_{33} + k_{43}}{6}.
\end{aligned}$$

As an example we apply the above to the first order system $y_1'(x) = y_2$, $y_2'(x) = y_1$ with initial values $y_1(0) = 0$, $y_2(0) = 1$. The numerical results are given in Table 7.6, below.

x_k	$y_1(x_k)$	$y_2(x_k)$
0.0	0.	1.
0.1	0.1001666666	1.00500416
0.2	0.2013359032	1.04534124
0.3	0.3045204293	1.04534124
0.4	0.4107529748	1.08107652
0.5	0.5210967534	1.12763162
0.6	0.636656129	1.18547252
0.7	0.7585876686	1.25517813
0.8	0.8881117157	1.33744612
0.9	1.0265246	1.43309989
1.0	1.17521163	1.5430968

Table 7.6

Problem 7.12: Find the solution to the initial value problem in the last paragraph. If this is too difficult show that $y_1(x) = \frac{1}{2}e^t - \frac{1}{2}e^{-t}$ and $y_2(x) = \frac{1}{2}e^t + \frac{1}{2}e^{-t}$ is the solution to this problem.

We close this section with a few remarks on solving higher order initial value problems. Suppose we have the mth order initial value problem

$$y^{(m)}(x) = f(x, y, y', \ldots, y^{(m-1)})$$

with $y(a) = y_a$, $y'(a) = y_a'$, $\ldots$, $y^{(m-1)}(a) = y_a^{(m-1)}$. We may turn this into a system of equations by introducing the variables

$u_1(x) = y(x)$, $u_2(x) = y'(x)$, $\ldots$, $u_m(x) = y^{(m-1)}(x)$. Then we have the following first order initial value system

$$U'(x) = F(x,U) = \begin{bmatrix} u_2 \\ u_3 \\ \vdots \\ f(x, y, y', \ldots, y^{(m-1)}) \end{bmatrix} = \begin{bmatrix} u_2 \\ u_3 \\ \vdots \\ f(x, u_1, \ldots, u_m) \end{bmatrix}$$

with

$$U(a) = \begin{bmatrix} y_a \\ y'_a \\ \vdots \\ y_a^{(m-1)} \end{bmatrix},$$

since $u'_k(x) = (y^{(k-1)}(x))' = y^{(k)}(x) = u_{k+1}$ and $u_k(a) = y^{(k-1)}(a) = y_a^{(k-1)}$ for $k = 1, 2, \ldots, m$. For example, consider the third order initial value problem

$$y'''(x) - x^2 y''(x) + xy'(x) + e^x y(x) = e^{-x} \sin x\,,$$

with the initial values $y(0) = 3$, $y'(0) = 1$, $y''(0) = -1$. This is transformed into the first order sytem of equations

$$U'(x) = \begin{bmatrix} u'_1(x) \\ u'_2(x) \\ u'_3(x) \end{bmatrix} = \begin{bmatrix} u_2(x) \\ u_3(x) \\ e^{-x} \sin x - e^x u_1(x) - x u_2(x) + x^2 u_3(x) \end{bmatrix}$$

with the initial vlaues

$$U(0) = \begin{bmatrix} 3 \\ 1 \\ -1 \end{bmatrix}.$$

Problem 7.13: Solve $y'' - y = 0$; $y(0) = 1$, $y'(0) = 1$ by using the method of the last paragraph. Does your result look familiar?

We note that a similar technique will work for systems of higher order initial value problems.

Exercise Set 7

1. This problem will prove the existence and uniqueness of the solution to the initial value problem

$$y'(x) = xy(x)\,, \quad y(0) = 1\,.$$

We first prove the existence of a solution

a) Define the sequence of functions $\{\varphi_n(x)\}$ by $\varphi_0(x) = 1$ and for $n \geq 0$,

$$\varphi_{n+1}(x) = \int_0^x t\,\varphi_n(t)\;dt + 1\,.$$

Show that $\varphi_n(0) = 1$ for all values of n. Note that this is essentially an integrated version of the differential equation with the correct boundary condition, that is, if $\varphi(x)$ is a solution of the initial value problem, then

$$\varphi(x) = \int_0^x t\,\varphi(t)\;dt + 1\,.$$

b) Start with $\varphi_0(x) \equiv 1$ and show that the sequence $\{\varphi_n(x)\}$ is a sequence of polynomials in x. Use mathematical induction to prove a formula for $\varphi_n(x)$.

c) Use the ratio test to prove that the sequence determined in (b) converges to a function defined by a power series for all values of x. This finishes the proof of existence.

Secondly, we prove the uniqueness of the solution

d) Let $\varphi(x)$ and $\psi(x)$ be two solutions of the initial value problem. Show that

$$\varphi(x) - \psi(x) = \int_0^x t\big(\varphi(t) - \psi(t)\big)\, dt\,.$$

e) If we restrict x to lie in the interval $[0, A]$, show that

$$|\varphi(x) - \psi(x)| \leq A \int_0^x |\varphi(x) - \psi(x)|\, dt\,.$$

f) Let

$$u(t) = \int_0^x |\varphi(t) - \psi(t)|\, dt\,.$$

Show that $u(0) = 0$ and $u(x) \geq 0$ for all x. Use the result of (e) to show that

$$u'(x) = -Au(x) \leq 0$$

for all $x \geq 0$.

g) Use the integrative factor e^{-Ax} to show that

$$e^{-Ax}u(t) \leq 0$$

for all $x \geq 0$. (Recall that $e^{-Ax} > 0$ for all values of x.)

h) Conclude from (f) and (g) that $u(x) \equiv 0$ for all values of x, and so, the two solutions $\varphi(x)$ and $\psi(x)$ are indeed the same.

2. Repeat Exercise 1 for the initial value problem

$$y'(x) = 3x^2(y^2(x) + 1), \quad y(0) = 0.$$

3. Suppose we wish to evaluate the integral

$$F(x) = \int_0^x \frac{dt}{\sqrt{1+t^3}},$$

where x is allowed to vary over the interval $[0, a]$, for some value of $a > 0$. Turn this into an intial value ordinary differential equation.

4. Repeat Exercise 3 with the integral

$$F(x) = \int_1^x \frac{dt}{\log t}.$$

5. Show that the Richardson errors for the formula

$$y_h^R(x) = 2y_{h/2}(x) - y_h(x)$$

has error $O(h^2)$.

6. Derive the formula for $y_h^R(x)$ if $y_h(x)$ is given in (7.15). Show that the associated error is $O(h^3)$.

7. Write up an algorithm for a fourth order Taylor series algorithm for the general first order initial value problem

$$y'(t) = f(x, y) \quad y(t_0) = x_0.$$

How many function evaluations will be necessary?

8. Do the calculations to derive the fourth order Runge-Kutta algorithm of Section 7.2 and establish the estimate for the local truncation error.

9. Derive the Adams-Moulton two-step implicit method with global error $O(h^3)$.

10. Write the following

$$x^2y''(x) + xy'(x) + 2y(x) = e^{-x}, \quad y(0) = 1, \quad y'(0) = -1$$

as a first order initial value problem involving systems.

11. Indicate how one might go about solving the following system of second order equations

$$\begin{gathered} y_1''(x) + y_1'(x) = y_1(x)y_2(x) \\ y_2''(x) + xy_2'(x) = (x^2+1)y_1(x)(y_2(x)+1) \text{ with} \\ y_1(0) = 1, \quad y_1'(0) = 1, \quad y_2(0) = 1, \quad y_2'(0) = -1. \end{gathered}$$

12. Show that in the multistep method given by (7.18) of Section 7.3, $\alpha_1 + \cdots + \alpha_k = 1$. (HINT: consider the equation $y' = 0$, $y(a) = y_a$.) Find a second relationship between the α's and β's. (HINT: consider the equation $y' = 1$, $y(0) = 0$.)

13. Derive the results given in (7.23) and (7.24) of Section 7.3.

14. Solve the initial problem

$$y'(x) = y(x) - e^{-x}, \quad y(0) = 0.5$$

using Euler's method with Richardson extrapolation to calculate the values of $x(t)$ on the interval $[0, 1]$. How small of a step size is necessary to get 4 place accuracy?

15. a) Use the algorithm of Thorem 7.2, the second order Runge-Kutta method, to calculate the solution of the initial value problem

$$y'(x) = y \cos x\,, \quad y(0) = 1$$

over the interval $[0, 1]$ with a step size that guarantees 4 place accuracy.

b) Repeat this problem with the second order Taylor Series method ((7.14) of Section 7.2).

16. Solve the initial value problem

$$y'(x) = y^{\frac{1}{3}}(x)\,, \quad y(0) = 8$$

using both Euler's method and the fourth order Runge-Kutta method on the interval $[0, 1]$. Use a step size of $h = 0.05$. Compare these results with the value of the exact solution.

17. a) State the third order Taylor series algorithm for the initial value problem

$$y'(x) = 3x^2y^2\,, \quad y(0) = 1\,.$$

b) Solve this equation numerically using the algorithm of part (a) for x on the interval $[0, 1]$. Select the step size to get 4 place accuracy.

18. Recall from Chapter 5 that we approximated the derivative as follows

$$y'(x) = \frac{-y(x-h) + y(x+h)}{2h}\,.$$

This suggests that to solve

$$y'(x) = f(x, y)\,, \quad y(t_0) = y_0\,,$$

that we try the algorithm

$$y_{n+1} = y_{n-1} + 2hf(x_n, y_n),$$

where (x_0, y_0) is the initial condition. Try this algorithm on the equation

$$y'(x) = -y(x), \quad y(0) = 1.$$

Use a step size of $h = 0.05$ to calculate the values over the interval $[0, 1]$. Can you make this work? Is it worth it? Compare these results with the exact values.

19. Solve the initial value problem

$$y'(t) = 2x^2y^3, \quad y(0) = 1$$

on the interval $[0, 1]$ by using the Runge-Kutta method with a step size of $h = 0.05$. Compare these results with the exact values.

20. a) Derive a third order Runge-Kutta algorithm for the first order initial value problem

$$y'(x) = f(x, y), \quad y(x_0) = y_0.$$

    b) Write and test a subroutine that solves Exercise 10. Compare the answer obtained by the two algorithms.

21. a) Solve the initial value problem

$$y'(x) = 2x(y + 1), \quad y(0) = 0$$

at $x = 0.1$ by using a fourth-order Runge-Kutta method and a step size of $h = 0.005$.

b) Solve the initial value problem

$$y'(x) = 2x(y+1)\,, \quad y(0) = 0.5$$

at $t = 01$ by using a fourth order Runge-Kutta method and a step size of $h = 0.005$.

c) Compare the results of (a) and (b).

22. Consider the following modification of the fourth order Runge-Kutta method due to Gill. We write

$$y_{n+1} = y_n + \frac{1}{6}(k_1 + 2ak_2 + 2bk_3 + k_4)$$

where

$$\begin{aligned} k_1 &= hf(x_n, y_n) \\ k_2 &= hf(x_n + \frac{h}{2}, y_n + \frac{k_1}{2}) \\ k_3 &= hf(x_n + \frac{h}{2}, y_n + ck_1 + ak_2) \\ k_4 &= hf(x_n + h, y_n + dk_2 + bk_3) \end{aligned}$$

where

$$a = 1 - \frac{1}{\sqrt{2}},\; b = 1 + \frac{1}{\sqrt{2}}, c = -\frac{1}{2} + \frac{1}{\sqrt{2}} \text{ and } d = -\frac{1}{\sqrt{2}}.$$

The use of these coefficients is designed to minimize round-off error. The classical fourth order Runge-Kutta method given in the text uses the coefficients $a = b = 1$ and $c = d = 0$. Rerun Exercises 10 and 11 using this improvement and compare your results.

23. Solve the initial value problem

$$y'(x) = \sin xy\,, \quad y(1) = 1$$

using the fourth order predictor-corrector algorithm on the interval $[0, 1]$ with a step size of $h = 0.05$. Then solve the same problem with the Runge-Kutta algorithm and the fourth order Taylor series algorithms using a step size of $h = 0.05$ in each case. Finally, compare the results of the three runs.

24. a) Solve the initial value problem

$$y'(x) = 10y(x) + 11x - 5x^2 - 1\,, \quad y(0) = 0$$

over the interval $[0, 1]$ using the fourth order Runge-Kutta method and a step size of $h = 0.05$.

b) Repeat part (a), but with the initial value condition $y(0) = 0.01$. How wide a discrepancy is there in the two solutions?

c) Explain this discrepency.

25. Solve the initial value problem

$$y'(x) = 100(\sin x - y(x))\,, \quad y(0) = 0$$

over the interval $[0, 1]$ for $h = \frac{1}{8}, \frac{1}{16}, \frac{1}{32}$, and $\frac{1}{128}$ using the fourth order Runge-Kutta method.

26. Use a fourth order Runge-Kutta algorithm to evaluate the integral in Exercise 4. That is, calculate $F(11)$. Use a step size of $h = 0.05$.

27. Solve the first order system

$$y_1'(x) = y_1(x) + xy_2(x)\,, \quad y_1(0) = 1$$
$$y_2'(x) = y_2(x) + x^2y_1(x)\,, \quad y_2(0) = 1$$

using the fourth order Runge-Kutta method with a step size of $h = 0.05$.

28. Solve the system

$$y_1'(x) = y_1(x)(y_1(x) - y_2(x))\,, \quad y_1(0) = 1$$
$$y_2'(x) = y_2(x)(y_1(x) - y_2(x))\,, \quad y_2(0) = 1$$

using a fourth order Runge-Kutta algorithm over the interval $[0, 2]$ with a step size of $h = 0.05$.

29. Solve the following "predator-prey" system

$$y_1'(x) = 2y_1(x)(y_2(x) + 3)\,, \quad y_1(0) = 2$$
$$y_2'(x) = -y_2(x)(y_1(x) - 1)\,, \quad y_2(0) = 1\,,$$

using the fourth order Runge-Kutta algorithm over the interval $[0, 2]$ with a step size of $h = 0.05$.

30. Solve the second order equation

$$(x^2 + 1)y''(x) + 2xy'(x) - (1 - x^2)y(x) = 0\,, \text{ where}$$
$$y(0) = 1\,, \quad y'(0) = 1\,.$$

using a fourth order predictor-corrector method with a step size of $h = 0.05$.

31. Solve

$$y_1'(x) = ay_1(x) - ry_2(x)$$
$$y_2'(x) = by_2(x) - sy_3(x)$$
$$y_3'(x) = cy_3(x) - uy_1(x)$$

with $a = 21$, $b = 2$, $c = b$, $r = 9$, $s = 4$, $u = 7$ and initial values $x_1(0) = 300$, $x_2(0) = 598$, $x_3(0) = 323$, over the interval $[0, 15]$ with a step size of 5 with the Runge-Kutta algorithm. This problem arises in a three species predator-prey model.

8 *Partial Differential Equations*

8.1 Introduction

In this chapter we study numerical methods for solving partial differential equations. These equations are our fundamental tool in describing the physical world. We will soon see that for many reasons the study of partial differential equations is much more difficult than the study of ordinary differential equations. The reader may recall that in Section 7.2 we gave a gernal result on the existence and uniqueness of solutions for first order initial value ordinary differential equations. The point we wish to emphasize is that for the general partial differential equation such results do not exist.

As an example of nonuniqueness consider the partial differential equation

$$u_{xx} - u_{yy} = 0\,. \tag{8.1}$$

Let f and g be two functions that possess second derivatives and let

$$w(x, y) = f(x + y) + g(x - y)\,.$$

Then, by the chain rule for partial derivatives,

$$w_{xx} = f''(x+y) + g''(x-y)$$

and

$$w_{yy} = f''(x+y) + g''(x-y)\,,$$

so that

$$w_{xx} - w_{yy} = 0\,.$$

In the case of ordinary differential equations uniqueness was roughly determined up to a constant. Thus, even in this simple case of the partial differential equation (8.1) we are nowhere near this kind of uniqueness since any function w as given above satisfies (8.1).

In this chapter we shall be interested in studying numerical methods for solving the general second order partial differential equation

$$Au_{xx} + Bu_{xy} + Cu_{yy} + Du_x + Eu_y + Fu + G = 0\,, \tag{8.2}$$

where A, B, C, D, E, F, and G are given functions of x and y, which are continuous in some region R of the xy-plane. Note that we have not specified any data conditions. This is another place where partial differential equations differ from ordinary differential equations: the given data must be specified in accordance with the class of equations to which equation (8.2) belongs.

For both practical and theoretical reasons the partial differential equation (8.2) is classified in one of three ways:

1) If $B^2 - 4AC < 0$ throughout R, then the equation is said to be *elliptic*.

As an example of this class of equations we have the Helmholz equation

$$u_{xx} + u_{yy} + fu = g\,, \quad a \le x \le b\,, \quad c \le y \le d\,, \tag{8.3}$$

where f and g are functions continuous on the region. Here we specify the values of u on the boundary. That is, we must specify

$$u(x,c) \text{ and } u(x,d) \text{ for } a \le x \le b$$

and

$$u(a,y) \text{ and } u(b,y) \text{ for } c \le y \le d\,.$$

A physical problem for which this would be the mathematical model would be steady state temperature distribution in the rectangle. The boundary conditions give steady state conditions for the temperatures on the boundary.

2) If $B^2 - 4AC = 0$ throughout R, then the equation is said to be *parabolic.*

As an example we have one version of the heat equation where we are given a rod with initial temperature distribution $f(x)$ on the interval $a \le x \le b$ and we apply heat to the ends of the rod over time. We are interested in the temperature distribution $u(x,t)$ in the rod over time. The relevant equations are

$$\begin{aligned} &u_{xx} = \alpha^2 u_t\,, \quad a \le x \le b\,, \quad t \ge 0 \\ &u(x,0) = f(x)\,, \quad a \le x \le b \\ &u(a,t) = g_1(t)\,, \quad t \ge 0 \\ &u(b,t) = g_2(t)\,, \quad t \ge 0 \end{aligned} \tag{8.4}$$

where α is a constant that depends on the material that makes up the rod.

3) If $B^2 - 4AC > 0$ throughout R, then the equation is said to be *hyperbolic.*

As an example, we have one version of the wave equation where we are given a string along the x-axis over the interval $a \le x \le b$. The string has an initial displacement $f(x)$, and an initial velocity $g(x)$.

If we require that the ends of the string are fixed, then we will have conditions at the endpoints that hold throughout time. The relevant equations are, for example

$$(8.5)\qquad \begin{aligned} &u_{xx} = \alpha^2 u_{tt}\,, \quad a \le x \le b\,, \quad t \ge 0 \\ &u(x,0) = f(x)\,, \quad a \le x \le b \\ &u_t(x,0) = g(x)\,, \quad a \le x \le b \\ &u(a,t) = u(b,t) = 0\,, \quad t \ge 0\,, \end{aligned}$$

where α is a constant that depends on the material of the string. The last condition requires that the endpoints stay on the x-axis. Since $u(x,t)$ represents the displacement off the x-axis at the point x and time t, if we wanted the ends off the x-axis, but fixed, then we would specify other constant values for $u(a,t)$ and $u(b,t)$. There are other types of conditions that allow the ends to vary. They sometimes involve free boundary conditions and involve the partials u_x (see Exercise 8.7).

It might be noted that if A, B and C are indeed functions which can vary over the region R, then it would be possible for an equation to change its classification inside the region R. This sort of behavior never occurs with ordinary differential equations and is too complex to be treated here.

We shall consider each of these three classes of equations in the following three sections. We shall apply the same general method to each class of equation and we shall see that each class possesses its own idiosyncracies. The method is that of difference equations. Recall, from Section 6.1, that we have the following numerical formulas for first and second derivatives of functions of one independent variable. They are, for the first derivative,

$$(8.6)\qquad f'(x) \approx \frac{f(x+h) - f(x)}{h}\,,$$

$$(8.7)\qquad f'(x) \approx \frac{f(x) - f(x-h)}{h}$$

and

$$f'(x) \approx \frac{f(x+h) - f(x-h)}{2h}. \tag{8.8}$$

For the second derivative we use the central difference formula

$$f''(x) \approx \frac{f(x+h) - 2f(x) - f(x-h)}{h^2}. \tag{8.9}$$

We shall now use each of these types of formulas for various classes of partial differential equations. For example, the wave equation (8.5) will be turned into the difference equation

$$\begin{aligned} &\frac{u(x+h,t) - 2u(x,t) + u(x-h,t)}{h^2} \\ &\qquad = \alpha^2 \, \frac{u(x,t+k) - 2u(x,t) + u(x,t-k)}{k^2}, \end{aligned} \tag{8.10}$$

where h and k are the horizontal and vertical step sizes. The problem is then to tie in the difference equation (8.10) with the boundary data. This is done in Section 8.4.

It is instructive to note that the points needed to approximate (8.5) are in the five point form or arrangement

$$\begin{array}{ccc} & (x, y+k) & \\ & \bullet & \\ \bullet & \bullet & \bullet \\ (x-h, y) & (x, y) & (x+h, y) \\ & \bullet & \\ & (x, y-k) & \end{array}$$

This introduces a grid system on the region R and for this reason all of the regions we will consider are rectangular. If the region is not rectangular or such that the grid points lie on the boundary, then we must do some sort of interpolation, which will complicate matters. One possible procedure is to change the step sizes near the boundary in order to put grid points on the boundary. This will work, but the error analysis near the boundary becomes very complicated and would be considerably different from the error estimates in the interior.

There are other, higher order difference methods that use more points. These methods work in the same way as the ones we will present, except that the difference equation obtained will be more complicated. For example a sixth order formula is known as the 9-point formula and uses the points displayed above plus the other four corner points of the implied square.

8.2 Elliptic Equations

In this section we will see how difference methods can be applied to elliptic differential equations. Recall that for an elliptic problem we must specify the values of the solution $u(x, y)$ on the entire boundary. To illustrate the general method we shall consider the elliptic

$$u_{xx} + u_{yy} + fu = g\,, \tag{8.11}$$

where $f(x, y)$ and $g(x, y)$ are given functions that are continuous on the boundary of our given region R. In order to amplify the presentation we shall assume that R is a rectangle,

$$\{(x, y) : a \leq x \leq b\,, c \leq y \leq d\}.$$

If we space our grid points over the rectangle $[a, b] \times [c, d]$ with step size h in the x variable and step size k in the y variable, then we determine the number of grid points by defining n and m by

$$nh = b - a \quad \text{and} \quad mk = d - c\,.$$

If we let $x_i = a + ih$, $i = 0, 1, \ldots, n$ and $y_j = c + jk$, $j = 0, 1, \ldots, m$, then we will denote by $w_{i,j}$ the calculated value of u at the grid points (x_i, y_j). Recall that we know the values of u on the boundary of our rectangle. Thus in terms of the w's this means that we know the values of $w_{0,j}$, $w_{n,j}$, $j = 0, \ldots, m$, and $w_{i,0}$, $w_{i,m}$, $i = 0, \ldots, n$. As a difference equation in the w's equation (8.11) becomes, using the equation (8.9) of the introduction,

$$\begin{aligned} &\frac{w_{i+1,j} - 2w_{i,j} + w_{i-1,j}}{h^2} + \frac{w_{i,j+1} - 2w_{i,j} + w_{i,j-1}}{k^2} \\ &\qquad + f_{i,j} w_{i,j} = g_{i,j} \end{aligned} \tag{8.12}$$

where $f_{i,j} = f(x_i, y_j)$ and $g_{i,j} = g(x_i, y_j)$. If we multiply both sides of (8.12) by h^2k^2 and collect the terms we have

$$\begin{aligned} k^2 w_{i+1,j} - \left(2h^2 + 2k^2 - h^2k^2 f_{i,j}\right) w_{i,j} + k^2 w_{i-1,j} \\ + h^2 w_{i,j+1} + h^2 w_{i,j-1} = h^2k^2 g_{i,j} \,. \end{aligned}$$

Finally we let $\lambda = \frac{h^2}{k^2}$ and divide by $-k^2$. This gives us our final difference equation,

$$\begin{aligned} - w_{i+1,j} + (2 + 2\lambda - h^2 f_{i,j}) w_{i,j} - w_{i-1,j} \\ - \lambda w_{i,j+1} - \lambda w_{i,j-1} = -h^2 g_{i,j} \,, \end{aligned} \tag{8.13}$$

where $i = 1, \ldots, n-1$ and $j = 1, \ldots, m-1$.

If we were to write this equation in matrix form we would have quite a few zero entries since we have $(n-1)(m-1)$ unknowns and only five of them appear in any one equation. To make this matrix a little less sparse we relabel our unknowns from left to right and up. This is called the *natural ordering* and is accomplished by introducing the new index

$$N = i + (j - i)(n - 1)$$

for $1 \leq i \leq n-1$ and $1 \leq j \leq m-1$. By doing this we put the coefficients $2 + 2\lambda - h^2 f_{i,j}$ on the main diagonal.

Problem 8.1: Justify the statement in the last sentence.

To illustrate these equations in a concrete setting let us take $a = 0$, $b = 1$, $c = 0$, $d = 1$, $h = \frac{1}{3}$ and $k = \frac{1}{4}$. Then $n = 3$ and $m = 4$. This gives us the following system of equations:

$$
\begin{aligned}
- w_{21} + (2 + 2\lambda - h^2 f_{11})w_{11} - w_{01} - w_{12} - w_{10} &= -h^2 g_{11} \\
- w_{31} + (2 + 2\lambda - h^2 f_{21})w_{21} - w_{11} - w_{22} - w_{20} &= -h^2 g_{21} \\
- w_{22} + (2 + 2\lambda - h^2 f_{12})w_{12} - w_{02} - w_{13} - w_{11} &= -h^2 g_{12} \\
- w_{32} + (2 + 2\lambda - h^2 f_{22})w_{22} - w_{12} - w_{23} - w_{21} &= -h^2 g_{22} \\
- w_{23} + (2 + 2\lambda - h^2 f_{13})w_{13} - w_{03} - w_{14} - w_{12} &= -h^2 g_{13} \\
- w_{33} + (2 + 2\lambda - h^2 f_{23})w_{23} - w_{13} - w_{24} - w_{22} &= -h^2 g_{23} \,.
\end{aligned}
$$

At first glance it may appear that we have more than $(n-1)(m-1) = 2\times 3 = 6$ unknowns in these 6 equations. If we recall that any w with a 0 or 3 as a first index or a 0 or 4 as a second index is a boundary value, we see that all of these quantities are known. Thus, under the natural ordering described above, we have the matrix equation

$$Ax = b \,,$$

where

$$
A = \begin{bmatrix}
2+2\lambda-h^2 f_{11} & -1 & -\lambda & 0 & 0 & 0 \\
-1 & 2+2\lambda-h^2 f_{21} & 0 & -\lambda & 0 & 0 \\
-\lambda & 0 & 2+2\lambda-h^2 f_{12} & -1 & -\lambda & 0 \\
0 & -\lambda & -1 & 2+2\lambda-h^2 f_{22} & 0 & -\lambda \\
0 & 0 & -\lambda & 0 & 2+2\lambda-h^2 f_{13} & -1 \\
0 & 0 & 0 & -\lambda & -1 & 2+2\lambda-h^2 f_{23}
\end{bmatrix}
$$

and

$$b = \begin{bmatrix} -h^2 g_{11} + w_{01} + \lambda w_{10} \\ -h^2 g_{21} + w_{31} + \lambda w_{20} \\ -h^2 g_{21} + w_{02} \\ -h^2 g_{13} + w_{03} + \lambda w_{14} \\ -h^2 g_{23} + w_{33} + \lambda w_{24} \end{bmatrix}.$$

Though it might not be obvious from so small an example, the general form of the coefficient matrix is a three banded matrix with the diagonal entries $2 + 2\lambda - h^2 f_{i,j}$ and the upper and lower off-diagonal entries are either -1 or 0. Then farther out from the diagonal are diagonals with $-\lambda$ as the entries. The distance from the main diagonal depends on the values of n and m.

It should also be noted that for h small enough the sytem (8.13) is diagonally dominant when naturally ordered, since λ is positive. Thus, rather than try to solve by Gaussian elimination a system that has many zero coefficients we might be inclined to try an iterative method, say Gauss-Seidel iteration. This is even more reasonable when it is realized that all the equations look the same, that is,

$$w_{i,j} = \frac{w_{i+1,j} + w_{i-1,j} + \lambda w_{i,j+1} + \lambda w_{i,j-1} - h^2 g_{i,j}}{2 + 2\lambda - h^2 f_{i,j}}$$

which in an iteration scheme could be written

$$w_{i,j}^{(n+1)} = \frac{w_{i+1}^{(n)} + w_{i-1,j}^{(n+1)} + \lambda w_{i,j+1}^{(n)} + \lambda w_{i,j-1}^{(n+1)} - h^2 g_{i,j}}{2 + 2\lambda - h^2 f_{i,j}}. \tag{8.14}$$

We illustrate all of the above with the following example. Consider the partial differential equation

$$u_{xx} + u_{yy} - 2u = 0$$

over the square $[0, 1] \times [0, 1]$ with boundary values

$$u(0, y) = e^y, \quad u(1, y) = e^{1+y}, \quad 0 \le y \le 1$$

and

$$u(x,0) = e^x, \quad u(x,1) = e^{x+1}, \quad 0 \le x \le 1.$$

The exact solution is $u(x,y) = e^{x+y}$.

Problem 8.2: Show that the solution given in the last paragraph satisfies the partial differential equation and the boundary values.

To make the calculations easier we shall take $h = k = 0.2$, so that $\lambda = 1$. Then $n = m = 5$ so that we have 16 equations in 16 unknowns. Since the values on the boundary are given, we need only calculate the 16 values $u(ih, jh)$, $i, j = 1, 2, 3, 4$. The true solution is $u(ih, jh) = e^{(i+j)h}$, which is given in the third column of each table below.

We first find a computed solution by solving the system of equations that the difference method produces using standard Gaussian elimination. In the interests of space we will not give the equations, but we urge the reader to work them out (see Exercise 8.1). Our results are given in Table 8.1, below.

In the next table, we give the results obtained by applying Gauss-Seidel iteration to this system of equations. The iteration scheme is the same as (8.14), except that in this case $\lambda = 1$ and $2 + 2\lambda - h^2 f_{i,j} = 4 + 2h^2 = 4.008$. Since $f = -2$ in this example, we are guaranteed that the system is diagonally dominant no matter what the step size chosen, and so we may use any starting values we wish. Below we give the iterated solution after 50 iterations.

I	*J*	Computed Values	Exact Values	Error
1	1	1.4922381	1.49182467	4.1343E-04
2	1	1.822763	1.93311877	6.4423E-04
3	1	3.33625835	2.2255409	7.1745E-04
4	1	2.71884928	2.71828179	5.6749E-04
1	2	4.822766297	1.82211877	6.442E-04
2	2	2.22655193	2.2255409	1.011033E-03
3	2	2.71940304	2.71828179	1.454583E-03
4	2	3.3209889	3.32011688	8.7202E-04
1	3	2.22625831	2.2255409	7.1741E-04
2	3	2.71940303	2.71828179	1.12124E-03
3	3	3.32136542	3.32011688	1.24854E-03
4	3	4.05618254	4.05519989	9.8265E-03
1	4	2.71884923	3.32011688	5.6744E-04
2	4	3.32098887	3.32011688	8.7199E-04
3	4	4.05618252	4.05519989	9.8263E-04
4	4	4.95383818	4.95303234	8.058E-04

Table 8.1

I	J	Computed Values	Exact Values	Error
1	1	1.49223813	1.49182467	4.1346E-04
2	1	1.82276305	1.82211877	6.4428E-04
3	1	2.2262584	2.2255409	7.175E-04
4	1	2.71884929	2.71828179	5.675E-04
1	2	1.82276305	1.82211877	6.4428E-04
2	2	2.22655207	2.2255409	1.01117E-03
3	2	2.71940318	2.71828179	1.12139E-03
4	2	3.32098896	3.32011688	8.7208E-03
1	3	2.2262584	2.2255409	7.175E-04
2	3	2.71940318	2.71828179	1.12139E-03
3	3	3.32136558	3.32011688	1.2587E-03
4	3	4.05618262	4.05519989	9.8273E-04
1	4	2.71884929	2.71828179	5.675E-04
2	4	3.32098896	3.32011688	8.7208E-04
3	4	4.05618262	4.05519989	9.8273E-04
4	4	4.95383822	4.95303234	8.0588E-04

Table 8.2

8.3 Parabolic Equations

We now consider numerical methods for solving parabolic equations. To be concrete about this we shall consider the heat equation

$$
\begin{aligned}
&\alpha^2 u_{xx} = u_t\,, \quad a \le x \le b\,, \quad t \ge 0 \\
&u(a,t) = g_1(t) \ \text{ and } \ u(b,t) = g_2(t) \ \text{ for } \ t \ge 0\,, \\
&u(x,0) = f(x) \ \text{ for } \ a \le x \le b
\end{aligned}
\tag{8.15}
$$

where $f(a) = g_1(0)$ and $f(b) = g_2(0)$.

In the xt plane the picture would be a semi-infinite strip with $a \le x \le b$ and $t \ge 0$ as pictured below.

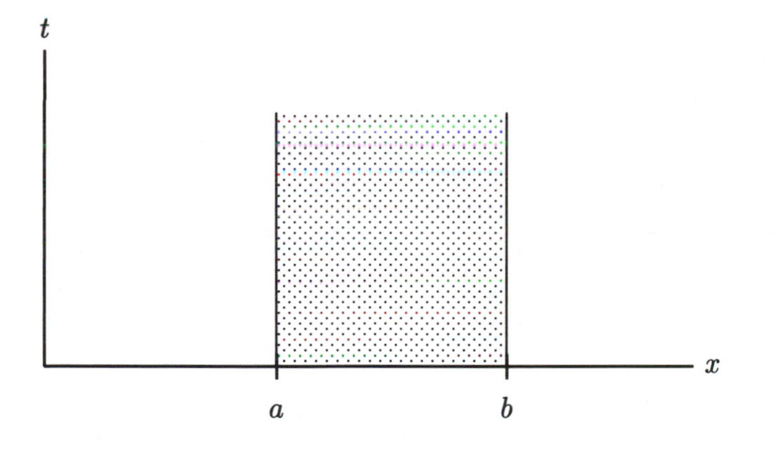

Figure 8.1

We solve this by the difference method used in the solution to the elliptic equation. We use formulas (8.6), (8.7) and (8.9) from Section 8.1 to set up our grid system. In this case, there were two different methods depending on whether formula (8.8) or (8.9) is used for the first derivative.

The general idea, in either case, is to get a four point grid system:

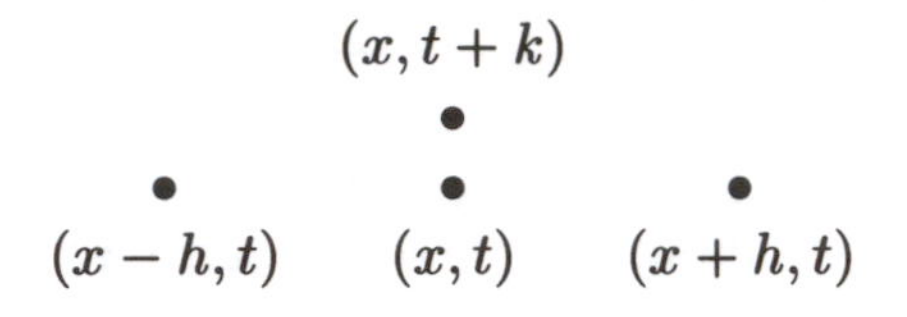

Forward differences

or

$(x-h, t)$ (x, t) $(x+h, t)$

$(x, t-k)$

Backward differences

The procedure, for a given value of x, is to compute the values of $u(x, t)$ for progressive values of t. However, as can be seen, we cannot march straight up as the diagram fans out. Instead, we must find all the values for a given t.

We define n by $nh = b - a$. We wish to end our computations at $t = T$, then we can define m by $mk = T$. Let $w_{i,j}$ denote the computed values of $u(a + ih, jk)$, $i = 0, 1, \ldots, n$, $j = 0, 1, \ldots, m$. Since we know the boundary values, we know exact values for $w_{0,j}$ and $w_{n,j}$, $j = 0, 1, \ldots, m$ and $w_{i,0}$, $i = 0, 1, \ldots, n$.

If we substitute the difference formulas for derivatives, as we have above in (8.12) of Section 8.2 we obtain the difference equation

$$\alpha^2 \frac{w_{i+1,j} - 2w_{i,j} + w_{i-1,j}}{h^2} = \frac{w_{i,j+1} - w_{i,j}}{k}. \tag{8.16}$$

Recall that we are interested in obtaining the values of u for higher t values, which means, in terms of the difference equation, for the next j value. Let $\lambda = \alpha^2 k/h^2$. Then we can solve for $w_{i,j+1}$ to get

$$w_{i,j+1} = \lambda w_{i+1,j} + (1 - 2\lambda) w_{i,j} + \lambda w_{i-1,j} \tag{8.17}$$

for $i = 1, \ldots, n-1$ and $j = 0, 1, \ldots, m$. This allows us to explicitly solve for the $j+1$ level if we know the values at the j level. Since we know the values at $j = 0$ and $j = m$ we can find the values for all j using the recursion (8.17). For this reason this method is said to be an *explicit* method. This method is pictured by the "forward differences" part of the last figure.

It can be shown that the method is stable if $\lambda < \frac{1}{2}$. *Stability*, in this context, means that the other solutions of the recursion relation (8.17), besides the "correct one" which is the approximation to the original equation (8.15), will not perturb the calculation greatly. If $\lambda > \frac{1}{2}$, then the errors at each step will be magnified at subsequent steps. This means one must choose h and k in such a way that

$$k \leq \alpha^2 \frac{h^2}{2}.$$

Thus, if h is small, then k is quite small, in fact, it might become too small, that is, too much computation will be necessary to advance the value of t very much. For example, if $\alpha = 1$, and $h = 0.01$, which is a reasonable value, then $k \leq 0.00005$, which is not so reasonable if you are interested in the value of $u(x, 1)$.

To combat this we can turn to an implicit method due to Crank and Nicolson. This difference method uses the backward difference for the derivative, that is, equation (8.7) of Section 8.1. If we use this form, the basic difference equation (8.16) becomes

$$\alpha^2 \, \frac{w_{i+1,j} - 2w_{i,j} + w_{i-1,j}}{h^2} = \frac{w_{i,j} - w_{i,j-1}}{k}.$$

Here we solve for $w_{i,j-1}$ and work backwards as suggested by the "backwards differences" part of the last figure. This gives us the recursion relation

$$w_{i,j-1} = -\lambda w_{i+1,j} + (1 + 2\lambda) w_{i,j} - \lambda w_{i-1,j}, \tag{8.18}$$

for $i = 1, \ldots, n-1$ and $j = 1, \ldots, m$.

If we put this recursion in matrix form we find

$$\begin{bmatrix} 1+2\lambda & \lambda & & & \\ \lambda & 1+2\lambda & -\lambda & & \\ & & \ddots & & \\ & & & -\lambda & 1+2\lambda \end{bmatrix} \begin{bmatrix} w_{1,j} \\ \vdots \\ w_{i,j} \\ \vdots \\ w_{n,j} \end{bmatrix} = \begin{bmatrix} w_{1,j-1} \\ \vdots \\ w_{i,j-1} \\ \vdots \\ w_{n,j-1} \end{bmatrix}$$

which is a diagonally dominant system since $1 + 2\lambda > 2\lambda$. Thus any iterative technique would work or we could just solve the tridiagonal system as it stands.

This is the advantage of the Crank-Nicolson algorithm: it remains stable for any choice of h and k. The major disadvantage is that we now have to solve a system of equations, but since we know we could solve this iteratively if desired, this is not such a big drawback.

As an example, consider the partial differential equation

$$4u_{xx} = u_t\,, \quad 0 \le x \le 1\,, \quad t \ge 0\,,$$

with

$$u(0,t) = e^{-\pi^2 t} \quad \text{and} \quad u(1,t) = 0 \quad \text{for} \quad t \ge 0$$

and

$$u(x,0) = \cos\frac{\pi x}{2} \quad \text{for} \quad 0 \le x \le 1\,.$$

The exact solution of this equation is $u(x,t) = e^{-\pi^2 t}\cos\frac{\pi x}{2}$.

Problem 8.3: Show that the solution given in the last paragraph satisfies the partial differential equation and the boundary values.

To illustrate the difference in the methods we shall give the results of running the forward difference algorithm with $h = 0.1$ and $k = 0.005$, so that $\lambda = \frac{1}{2}$ (see Table 8.3, below), and with $h = 0.1$ and $k = 0.01$, so that $\lambda = 1$ (see Table 8.4, below). In both cases we compute the value of $u(x,1)$ for $x = ih$, $i = 0, \ldots, 10$.

x	Computed Values	Exact Values	Error
0.0	0.3727078379	0.3727078379	0
0.1	0.36800538	0.3681191894	-1.138094E-04
0.2	0.3542670759	0.3544662192	-1.991433E-04
0.3	0.3318299306	0.3320851169	-2.823817E-04
0.4	0.3012445937	0.3015269754	-2.823817E-04
0.5	0.2632618598	0.2635442391	-2.823793E-04
0.6	0.2188142518	0.2190721704	-2.579186E-04
0.7	0.1689931303	0.1692058169	-2.126866E-04
0.8	0.1160218916	0.1151730557	-1.511641E-04
0.9	0.0582259156	0.0583043512	-7.84356E-05
1.0	0	0	0

Table 8.3

x	Computed Values	Exact Values	Error
0.0	0.3727078379	0.3727078379	0
0.1	2731404.34	0.3681191894	2731403.98
0.2	-4079196.45	0.3544662192	-4079196.8
0.3	3730612.59	0.3320851169	3730612.26
0.4	-2443456.73	0.3015269754	-2443457.03
0.5	1182810.32	0.2635442391	1182810.06
0.6	-418329.1849	0.2190721704	-418329.4039
0.7	103081.7883	0.1692058169	103081.6191
0.8	-15831.4505	0.1151730557	-15831.5656
0.9	4081.207658	0.0583043512	1081.149354
1.0	0	0	0

Table 8.4

As can be seen from the results, the computed values for $u(x, 1)$ in the second case, that is, $h = 0.1$ and $k = 0.01$, are totally incorrect as expected by our discussion of stability.

For the Crank-Nicolson algorithm we take $h = 0.1$ and $k = 0.01$, the values that did not give very good results in the forward difference algorithm. The following results show that this method gives quite good agreement to the exact values.

x	*Computed Values*	*Exact Values*	*Error*
0.0	0.3727078379	0.37270788379	0
0.1	0.36799826181	0.3681191894	-1.3657128E-04
0.2	0.3542073329	0.3544662192	-2.5888629E-04
0.3	0.3317916527	0.3320851169	-2.93464245E-04
0.4	0.301162703	0.3015269754	-3.64272393E-04
0.5	0.2631884412	0.2635442391	-3.55797918E-04
0.6	0.218734297	0.2190721704	-3.37873366E-04
0.7	0.1689569736	0.1692058169	-2.48843322E-04
0.8	0.1149886355	0.1151730557	-1.84420202E-04
0.9	0.0582110128	0.0583043512	-9.3338365E-05
1.0	0	0	0

Table 8.5

Since the formula used for the numerical approximate of the second derivative is second order, one might ask why we use a first order formula for the first derivative rather than the second order central difference formula. It can be shown that if we do this we will obtain a formula that is numerically unstable, that is the values obtained will diverge from the correct values.

8.4 Hyperbolic Equations

We conclude this chapter with a look at how the difference method can be applied to hyperbolic partial differential equations. The paradigm for these equations is the wave equation, which in one space variable has the form

$$
\begin{aligned}
&\alpha^2 u_{xx} = u_{tt}\,, \quad a \le x \le b\,, \quad t \ge 0 \\
&u(a,t) = u(b,t) = 0\,, \quad t \ge 0 \\
&u(x,0) = f(x)\,, \quad a \le x \le b \\
&u_t(x,0) = g(x)\,, \quad a \le x \le b\,.
\end{aligned}
\tag{8.19}
$$

Two remarks can be made with respect to the equation (8.19). First of all, this is the first equation where we have had to bring in information regarding the value of a derivative of the solution. Secondly, it is possible to write down, from the information given in (8.19), the analytical solution to (8.19). Indeed, the solution is

$$u(x,t) = \frac{1}{2}\big(f(x+\alpha t) + f(x-\alpha t)\big) + \frac{1}{2\alpha}\int_{x-\alpha t}^{x+\alpha t} g(s)\;ds\,.$$

It is easily verified that this is the solution (see Exercise 8.2), provided we assume that f is twice differentiable, g is integrable and we extend f over the entire real line by requiring that

$$f(-s) = -f(s) \;\; \text{and} \quad f(s + 2(b-a)) = f(s)\,.$$

As in the case of the heat equation of the previous section we are interested in advancing the values of $u(x,t)$ in terms of increasing t since we know the values where $t = 0$. We shall end up with a recursion relation much like that of (8.17) of the previous section.

We take h and k as the step sizes for our grid and define the integer n by $nh = b - a$. If we have a final value of t, say T, at which we wish to calculate $u(x,t)$, then we define the integer m by $mk = T$.

As before we denote by $w_{i,j}$ the calculated value of $u(a+ih, jk)$ and recall that the boundary conditions tell us that we know the values of $w_{i,0}$, $i = 0, 1, \ldots, n$ and $w_{0,j}$ and $w_{n,j}$ for $j = 0, 1, \ldots, m$. If we apply formula (8.9) of Section 8.1 to the partial differential equation we get

$$\alpha^2 \frac{w_{i+1,j} - 2w_{i,j} + w_{i-1,j}}{h^2} = \frac{w_{i,j+1} - 2w_{i,j} + w_{i,j-1}}{k^2}.$$

If we let $\lambda = \alpha^2 k^2/h^2$ we can rewrite this difference equation as the recursion relation

$$w_{i,j+1} = \lambda w_{i+1,j} + 2(1-\lambda)w_{i,j} + \lambda w_{i-1,j} - w_{i,j-1}, \tag{8.20}$$

for $i = 1, \ldots, n-1$ and $j = 1, \ldots, n$.

The recursion (8.20) works well except at the very beginning. While we know the values at $w_{i,0}$, namely $f(x_1)$, we don't know the values of $w_{i,1}$ and the recursion (8.20) only allows us to start with $w_{i,2}$. To get around this we approximate the condition $u_t(x, 0) = g(x)$. Since the formula (8.9) has order $O(h^2)$ we should approximate this starting value by an order two formula for the first derivative, namely equation (8.8) of Section 8.1

To do this note that $u_t(x, 0) = g(x)$ would approximate as

$$\frac{w_{i,1} - w_{i,-1}}{2k} = g(x_i).$$

To get rid of the $w_{i,-1}$ we use the recursion (8.20) with $j = 0$ and get

$$w_{i,1} = \lambda w_{i+1,0} + 2(1-\lambda)w_{i,0} + \lambda w_{i-1,0} - w_{i,-1}.$$

If we combine these two results we find that

$$\begin{aligned} w_{i,1} &= \lambda \frac{w_{i+1,0} + w_{i-1,0}}{2} + (1-\lambda)w_{i,0} + kg(x_i) \\ &= \lambda \frac{f(x_{i+1}) + f(x_{i-1})}{2} + (1-\lambda)f(x_i) + kg(x_i), \end{aligned} \tag{8.21}$$

since $w_{5,0} = u(x_5, 0) = f(x_5)$. Thus, the complete solution to the problem is given by the results (8.20) and (8.21).

It can be shown that this method is stable as long as $\lambda \le 1$. If $\lambda > 1$, then the errors made at one step would be magnified in subsequent steps. There do exist implicit algorithms, like the Crank-Nicolson algorithm for the heat equation, but we shall not discuss them here.

As an illustration, we consider the partial differential equation

$$\begin{aligned} &u_{xx} = u_{tt}\,, \quad 0 \le x \le 1\,, \quad t \ge 0 \\ &u(0,t) = u(1,t) = 0\,, \quad t \ge 0 \\ &u(x,0) = \sin(\pi x)\,, \quad 0 \le x \le 1 \\ &u_t(x,0) = \frac{\sin(2\pi x)}{4}\,, \quad 0 \le x \le 1\,. \end{aligned}$$

The exact solution to this equation is

$$u(x,t) = \sin \pi x \cos \pi t + \frac{\sin(2\pi x) sin(2\pi t)}{8\pi}\,.$$

For the computations in Table 8.6, below, we take $h = 0.1$ and $k = 0.05$ and use (8.20) and (8.21) to compute $u(x, 1)$.

Problem 8.4: Show that the solution given in the last paragraph satisfies the partial differential equation and the boundary values.

x	Computed Values	Exact Values	Error
0.0	0	0	0
0.1	-0.3108773525	-0.3090169969	-1.8603556E-03
0.2	-0.59079122	-0.5877852575	-3.0059625E-03
0.3	-0.81201257	-0.8090170036	-2.9955664E-03
0.4	-0.95288666	-0.9510565229	-1.8301371E-03
0.5	-0.99995298	-1	4.702E-05
0.6	-0.94913692	-0.95105652	1.9196E-03
0.7	-0.80594532	-0.809016996	3.071676E-03
0.8	-0.584724	-0.587785252	3.061252E-03
0.9	-0.3071275783	-0.309016994	1.8894157E-03
1.0	0	0	0

Table 8.6

Exercise Set 8

Mathematical Problems

1. Derive the system of equations needed to get the first set of solutions to the example of Section 8.2.

2. Verify that equation (8.20) of Section 8.4 is indeed the solution to equation (8.19) of that section under the conditions on f.

3. Establish error estimates for the difference methods described in the text for the various partial differential equations.

4. If

$$f(x) = a_0 + \sum_{n=1}^{N} (a_n \cos(n\pi x) + b_n \sin(n\pi x)),$$

then show that the solution to the heat equation

$$\begin{aligned} &u_{xx} = u_t, \quad 0 \le x \le 1, \quad t \ge 0, \\ &u(0,t) = u(1,t) = 0, \quad t \ge 0, \\ &u(x,0) = f(x), \quad 0 \le x \le 1 \end{aligned}$$

is given by

$$u(x,t) = a_0 + \sum_{n=1}^{N} e^{-(n\pi)^2 t} (a_n \cos(n\pi x) + b_n \sin(n\pi x)).$$

5. It can be shown that the general second order elliptic partial differential equation can be rewritten, by a change of variables, in the form

$$u_{xx} + a u_{yy} + b u_x + c u_y + d u = e, \quad \alpha \le x \le \beta, \quad \gamma \le y \le \delta,$$

where a, b, c, d and e are functions of x and y and a is a positive function. Write up the algorithm to use the difference methods of the text to solve this more general equation. Use the boundary conditions

$$\left.\begin{aligned} A_1 u(\alpha, y) + B_1 u_x(\alpha, y) = F_1(y) \\ A_2 u(\beta, y) + B_2 u_x(\beta, y) = F_2(y) \end{aligned}\right\} \gamma \le y \le \delta,$$

$$\left.\begin{aligned} A_3 u(x, \gamma) + B_3 u_y(x, \gamma) = G_1(x) \\ A_4 u(x, \delta) + B_4 u_y(x, \delta) = G_2(x) \end{aligned}\right\} \alpha \le x \le \beta,$$

where A_i, B_i, $i = 1, 2, 3, 4$ are constants.

6. The general second order parabolic partial differential equation can be rewritten, by a change of variables, in the form

$$u_{xx} + au_t + bu_x + cu = d\,, \quad \alpha \le x \le \beta\,, \quad t \ge 0\,,$$

where a, b, c and d are functions of x and t. Write up an algorithm, using the difference methods of the text, to solve this equation. Use the boundary conditions

$$\left.\begin{aligned} u(xl,0) &= f(x)\,, \quad \alpha \le x \le \beta\,, \\ A_1 u(\alpha,t) + B_1 u_x(\alpha,t) &= F_1(t)\,, \\ A_2 u(\beta,t) + B_2 u_x(\beta,t) &= F_2(t)\,, \end{aligned}\right\} \quad t \ge 0\,,$$

where A_1, A_2, B_1 and B_2 are constants.

7. The general second order hyperbolic partial differential equation can be rewritten, by a change of variables, in the form

$$u_{xx} - au_{tt} + bu_x + cu_t + du = e\,, \quad \alpha \le x \le \beta\,, \quad t \ge 0\,,$$

where a, b, c, d and e are functions of x and t, and a is a positive function. Write up an algorithm, using the difference methods of the text, to solve this equation. Use the boundary conditions

$$\begin{aligned} &u(x,0) = f(x)\,, \quad u_t(x,0) = g(x)\,, \quad \alpha \le x \le \beta\,, \\ &A_1 u(\alpha,t) + B_1 u_x(\alpha,t) = F_1(t)\,, \quad t \ge 0\,, \\ &A_2 u(\beta,t) + B_2 u_x(\beta,t) = F_2(t)\,, \quad t \ge 0\,, \end{aligned}$$

where A_1, A_2, B_1 and B_2 are constants. If either A_1 or A_2 is equal to zero, then the equation is said to have a free boundary condition.

8. Show that the solution to

$$u_{tt} - \alpha^2 u_{xx} = F(x,t)\,, \quad a \le x \le b\,, \quad t \ge 0$$
$$u(x,0) = f(x)\,, \quad u_t(x,0) = g(x)$$
$$u(a,t) = u(b,t) = 0$$

is given by

$$u(x,t) = \frac{f(x+\alpha t) + f(x-\alpha t)}{2} + \frac{1}{2\alpha}\int_{x-\alpha t}^{x+\alpha t} g(x)\, ds$$
$$+ \frac{1}{2\alpha}\int_0^t \int_{x-\alpha(t-\eta)}^{x+\alpha(t-\eta)} f(\xi,\eta)\, d\xi\, d\eta\,,$$

subject to certain conditions on f and F. What are they?

9. As an example of how to expand these methods to more variables consider Laplace's equation in three dimensions

$$u_{xx} + u_{yy} + u_{zz} = 0$$

$a \le x \le b$, $c \le y \le d$ and $e \le z \le f$, with $u(x,y,z)$ given on the boundary of the rectangular solid. Use the difference method, as given in the text, to write up an algorithm for numerically solving this partial differential equation.

Computational Problems

10. Let $R = \{(x, y) : 0 \le x \le 1\,,\ 0 \le y \le 1\}$. Consider the equation

$$\begin{aligned} &u_{xx} + u_{yy} = 0 \ \text{ on } \ R \\ &u = x \text{ on the boundary of } R. \end{aligned}$$

Approximate the value of $u(\frac{1}{2}, \frac{1}{2})$ with $h = 0.1$, $k = 0.1$.

11. Use $h = k = 0.1$ to solve the following equation for all of the interior points

$$\begin{aligned} &u_{xx} + u_{yy} = 0 \quad 0 \le x \le 1 \quad 0 \le y \le 1 \\ &u(x, 0) = 0\,, \quad u(x, 1) = x\,, \quad 0 \le x \le 1 \\ &u(0, y) = 0\,, \quad u(1, y) = y\,, \quad 0 \le y \le 1\,. \end{aligned}$$

12. Solve the following equation, using $h = k = 0.5$ for all of the interior points

$$\begin{aligned} &u_{xx} + u_{yy} = -2\,, \quad 0 \le x \le 1\,, \quad 0 \le y \le 1 \\ &u(0, y) = y\,, \quad u(1, y) = \sin h(\pi) \sin(\pi y)\,, \quad 0 \le y \le 1 \\ &u(0, x) = u(x, 1) = x(1 - x)\,, \quad 0 \le x \le 1\,. \end{aligned}$$

13. Use $h = 0.2$, $k = 0.1$ to solve the following equation for all interior points

$$\begin{aligned} &u_{xx} + u_{yy} = (x^2 + y^2)e^{xy}\,, \quad 0 \le x \le 2\,, \quad 0 \le y \le 1\,, \\ &u(0, y) = 1\,, \quad u(2, y) = e^{2y}\,, \quad 0 \le y \le 1\,, \\ &u(x, 0) = 1\,, \quad u(x, 1) = e^{x} \quad 0 \le x \le 2\,. \end{aligned}$$

14. Use $h = 0.1$, $k = 0.6$ to solve the following equation for all interior points

$$\begin{aligned} &u_{xx} + u_{yy} = xy(x-2)(y-2)\,, \quad 0 \le x \le 2\,, \quad 0 \le y \le 2 \\ &u(x,0) = u(x,2) = 0\,, \quad 0 \le x \le 2 \\ &u(0,y) = 1\,, \quad u(2,y) = 0\,, \quad 0 \le y \le 2 \end{aligned}$$

15. Use the algorithm developed in Exercise 5 to solve the equation for all interior points using $h = 0.1$, $k = 0.5$

$$2u_{xx} + u_{yy} - u_x = 2\,, \quad 0 \le x \le 1\,, \quad 0 \le y \le 1$$

with $u(x,y) = 0$ on the boundary.

16. Solve for $u(x,1)$ in the equation

$$\begin{aligned} &u_{xx} = u_t\,, \quad 0 \le x \le 1\,, \quad t \ge 0 \\ &u(0,t) = u(1,t) = 0\,, \quad t \ge 0 \\ &u(x,0) = x(1-x)\,, \quad 0 \le x \le 1\,. \end{aligned}$$

Use $h = 0.1$ and $k = 0.005$ in the forward difference method and $h = 0.1$ and $k = 0.01$ in the Crank-Nicolson backward difference method. Compare the results.

17. Solve the equation

$$\begin{aligned} &u_{xx} = u_t\,, \quad 0 \le x \le 2\,, \quad t \ge 0 \\ &u(0,t) = u(2,t) = 0\,, \quad t \ge 0 \\ &u(x,0) = \sin(2\pi x)\,, \quad 0 \le x \le 2\,. \end{aligned}$$

Use $h = 0.1$, $k = 0.01$ in the Crank-Nicolson method to approximate $u(x,0.5)$. Use $h = 0.1$, $k = 0.005$ in both methods to calculate $u(x,0.5)$ and compare the values.

18. Solve the equation

$$\begin{aligned}
&u_{xx} = u_t\,, \quad 0 \le x \le 2\,, \quad t \ge 0 \\
&u(0,t) = u(2,t) = 0\,, \quad t \ge 0 \\
&u(x,0) = x(2-x)\,, \quad 0 \le x \le 2\,.
\end{aligned}$$

Use $h = 0.1$ and $k = 0.01$ in the Crank-Nicolson method to calculate $u(x, 0.5)$.

19. Use the algorithm developed in Exercise 6 to solve

$$\begin{aligned}
&u_{xx} - u_t = -2\,, \quad 0 \le x \le 1\,, \quad t \ge 0 \\
&u(0,t) = u(1,t) = 0\,, \\
&u(x,0) = \sin(\pi x) + x(1-x)\,.
\end{aligned}$$

Use $h = 0.1$, $k = 0.01$ in the Crank-Nicolson method to calculate $u(x, 0.25)$.

20. Solve the equation

$$\begin{aligned}
&u_{xx} = u_t\,, \quad -1 \le x \le 1\,, \quad t \ge 0 \\
&u(-1,t) = u(1,t) = 0\,, \quad t \ge 0 \\
&u(x,0) = |x| - 1\,, \quad u_t(x,0) = 0\,, \quad -1 \le x \le 1\,.
\end{aligned}$$

Use $h = 0.1$, $k = 0.01$ to compute $u(x, 1)$.

21. Solve the equation

$$\begin{aligned}
&u_{tt} - u_{xx} = 0\,, \quad 0 \le x \le 1\,, \quad t \ge 0 \\
&u(0,t) = \sin^2 t\,, \quad u(1,t) = 0\,, \quad t \ge 0 \\
&u(x,0) = u_t(x,0) = 0\,, \quad 0 \le x \le 1\,.
\end{aligned}$$

Use $h = 0.1$, $k = 0.01$ to compute $u(x, 1)$.

22. Solve the equation

$$\begin{aligned}
&u_{tt} = u_{xx}\,, \quad 0 \le x \le 1\,, \quad t \ge 0 \\
&u(0,t) = u(1,t) = 0\,, \quad t \ge 0 \\
&u(x,0) = \sin(2\pi x)\,, \quad 0 \le x \le 1 \\
&u_t(x,0) = 2\pi \sin(2\pi x)\,, \quad 0 \le x \le 1\,.
\end{aligned}$$

Use $h = k = 0.1$ to calculate $u(x, 0.5)$.

23. Solve the equation

$$\begin{aligned}
&u_{tt} = u_{xx}\,, \quad 0 \le x \le 1\,, \quad t \ge 0 \\
&u(0,t) = u(1,t) = 0\,, \quad t \ge 0 \\
&u(x,0) = \sin(\pi x)\,, \quad 0 \le x \le 1 \\
&u_t(x,0) = 0\,, \quad 0 \le x \le 1\,.
\end{aligned}$$

Use $h = 0.1$, $k = 0.05$ to calculate $u(x, 0.5)$.

24. Use the algorithm developed in Exercise 7 above to solve the following equations. Use $h = 0.1$, $k = 0.01$ to calculate $u(x, 1)$.

a)

$$\begin{aligned}
&u_{tt} - u_{xx} = 1 - x\,, \quad 0 \le x \le 1\,, \quad t \ge 0 \\
&u(0,t) = u(1,t) = 0\,, \quad t \ge 0 \\
&u(x,0) = x^2(1 - x)\,, \quad 0 \le x \le 1 \\
&u_t(x,0) = 0\,, \quad 0 \le x \le 1\,.
\end{aligned}$$

b)

$$\begin{aligned}
&u_{tt} - u_{xx} = 1 - x\,, \quad 0 \le x \le 1\,, \quad t \ge 0 \\
&u(x,0) = x^2(1 - x)\,, \quad 0 \le x \le 1 \\
&u_t(x,0) = 0\,, \quad 0 \le x \le 1 \\
&u_x(0,t) = u_x(1,t) = 0\,, \quad t \ge 0\,.
\end{aligned}$$

c)
$$u_{tt} - u_{xx} = 1 - x\,, \quad 0 \le x \le 1\,, \quad t \ge 0$$
$$u(x,0) = x^2(1-x)\,, \quad 0 \le x \le 1$$
$$u_t(x,0) = 0\,, \quad 0 \le x \le 1$$
$$u_x(0,t) = u(1,t) = 0\,, \quad t \ge 0\,.$$

9 *Orthogonal Polynomials and Their Applications*

In this chapter we discuss orthogonal polynomials and a few of their many applications. The first section discusses the general theory of orthogonal polynomials. There is more material covered in this section than we shall use in the sections that follow, but this material is of interest in itself. The second section is concerned with another method of numerical quadrature called Gaussian quadrature. This method of integration has many advantages, as we shall see. Section 9.3 takes up the problem of least squares approximation, which we first discussed in Section 5.3, but from another point of view. In this section we discuss both continuous and discrete approximations. Finally, in Section 9.4 we discuss a way of making better use of Taylor series approximations for numerical approximations of functions.

9.1 Orthogonal Polynomials

We begin this chapter with some of the basic results in the theory of orthogonal polynomials that will be of use in the following section. For more on the theorem of orthogonal polynomials see the books of Szegö [14] and Rainville [10].

Definition 9.1: *We say that a sequence of polynomials* $\{p_n(x)\}_{n=0}^{+\infty}$ *is* **simple** *if and only if*

1) *the degree of* $p_n(x)$ *is* n *for* $n = 0, 1, 2, \ldots$

and

2) *there is exactly one polynomial of each degree in the set.*

Theorem 9.1: *If* $q(x)$ *is a polynomial of degree* m *and* $\{p_n(x)\}$ *is a simple set of polynomials, then there exist constants* $c_1, c_2, \ldots, c_m$ *such that*

$$q(x) = \sum_{k=0}^{m} c_k p_k(x) .$$

This result is easily proved by mathematical induction.

Definition 9.2: *Let* $\{p_n(x)\}_{n=0}^{+\infty}$ *be a sequence of real, simple polynomials. Let* $a < b$ *and let* $w(x)$ *be positive on the interval* $[a, b]$. *We say that* $\{p_n(x)\}$ *is* **orthogonal** *over* $[a, b]$, *with respect to the weight function* $w(x)$, *if and only if*

$$\int_a^b p_n(x) p_m(x) w(x)\, dx = 0$$

whenever $m \neq n$.

It should be noted that since $w(x) > 0$ we see that

$$g_n = \int_a^b p_n^2(x) w(x)\, dx > 0 \tag{9.1}$$

If $g_n = 1$ for all $n = 0, 1, 2, \ldots$, then the orthogonal polynomials are said to be *orthonormal.* The constant g_n is all that differentiates one set of polynomials orthogonal with respect to $w(x)$ over $[a, b]$ from another.

Example 9.1: Let us take the simplest classical case, namely, $a = -1$, $b = 1$ and $w(x) = 1$, for all x in the interval $[-1, 1]$. If $p_0(x) = k_{00}$, then we have

$$\int_{-1}^{1} p_0^2(x)\ dx = 2k_{00}^2\,,$$

so let us take $k_{00} = 1$.

To construct $p_1(x) = k_{11}x + k_{10}$ we must satisfy

$$\int_{-1}^{1} p_1^2(x)p_0(x)\ dx = 0$$

and

$$\int_{-1}^{1} p_1^2(x)\ dx > 0\,.$$

Thus

$$k_{10} = 0\,.$$

We then have

$$\int_{-1}^{1} p_1^2(x)\ dx = k_{11}^2 \int_{-1}^{1} x^2\ dx = \frac{2k_{11}^2}{3}\,,$$

so we take $k_{11} = 1$ to amplify matters, and so

$$p_1(x) = x\,.$$

Finally, we shall compute $p_2(x) = k_{22}x^2 + k_{21}x + k_{20}$. Here we must satisfy the three conditions

1) $\int_{-1}^{1} p_2(x)p_0(x)\ dx = 0$,

2) $\int_{-1}^{1} p_2(x)p_1(x)\ dx = 0$

and

3) $\int_{-1}^{1} p_2^2(x)\ dx > 0\,.$

Since $p_0(x) = 1$ and $p_1(x) = x$, conditions (1) and (2) become

1) $\int_{-1}^{1}(k_{22}x^2 + k_{21}x + k_{20})\,dx = 0$

and

2) $\int_{-1}^{1} x(k_{22}x^2 + k_{21}x + k_{20})\,dx = 0\,.$

Thus (1) and (2) give us the equations

$$\frac{2k_{22}}{3} + 2k_{20} = 0$$

and

$$\frac{2k_{21}}{2} = 0\,.$$

Thus, we have $k_{21} = 0$ and $k_{20} = -k_{22}/3$. Thus

$$p_2(x) = k_{22}(x^2 - \frac{1}{3})\,.$$

Condition (3) gives us

$$0 < \int_{-1}^{1} p_2^2(x)\,dx = \frac{8k_{22}^2}{45}\,.$$

The value of k_{22} is thus completely arbitrary. The usual choice is $k_{22} = \frac{3}{2}$, so that

$$p_2(x) = \frac{1}{2}(3x^2 - 1)\,.$$

Thus we have the first three *Legendre polynomials*

$$p_0(x) = 1$$
$$p_1(x) = x$$

and

$$p_2(x) = \frac{1}{2}(3x^2 - 1)\,.$$

It might be noted these are even or odd polynomials depending on the index, that is, $p_{2k}(x)$ is an even polynomial and $p_{2k+1}(x)$ is an odd polynomial. (See Exercise 9.3.)

We list below some of the more common orthogonal polynomials with their weights and intervals.

Jacobi polynomials: $P_n^{(a,b)}(x)$.

Here the interval is $[-1,1]$ and the weight function is $w(x) = (1-x)^a(1+x)^b$, where $a, b > -1$. Special cases are given by choosing values for a and b. In particular, choosing

1) $a = b = 0$ yields the Legendre polynomials, $P_n(x)$,

and

2) $a = b = -\frac{1}{2}$ yields the Chebychev polynomials of the first kind, $T_n(x)$.

Laguerre polynomials: $L_n^{(a)}(x)$.

Here the interval is $[0, +\infty]$ and the weight function is $e^{-x}x^a$, where $a > -1$. Usually one writes $L_n(x)$ for $L_n^{(0)}(x)$.

Hermite polynomials: $H_n(x)$.

Here the interval is $(-\infty, \infty)$ and the weight function is e^{-x^2}.

The following result gives an equivalent condition for orthogonality and its proof, which we omit, follows easily from Definitions 9.1 and 9.2 and Theorem 9.1. (See Exercise 9.1.)

Theorem 9.2: *If $\{p_n(x)\}$ is a simple set of real polynomials and $w(x)$ is positive on the interval $[a,b]$, where $a < b$, then a necessary and sufficient condition for $\{p_n(x)\}$ to be orthogonal with respect to the weight $w(x)$ over the interval $[a,b]$ is that*

$$\int_a^b x^k p_n(x) w(x)\, dx = 0$$

for all k, $k = 0, 1, 2, \ldots, n-1$.

From this result we can deduce two useful corollaries.

Corollary 9.3: *If* $q(x)$ *is a polynomial of degree less than* n *and* $\{p(x)\}$ *is orthogonal over* $[a, b]$ *with respect to the weight* $w(x)$*, then*

$$\int_a^b q(x)p_n(x)w(x)\,dx = 0\,.$$

Corollary 9.4: *Under the hypothesis of Corollary 9.3 we have*

$$\int_a^b x^n p_n(x)w(x)\,dx > 0\,.$$

The following result is essential for some of the formulations involved in Gaussian integration.

Theorem 9.5: *If* $\{p_n(x)\}$ *is orthorgonal with respect to the weight* $w(x)$ *over the interval* $[a, b]$*, then*

1) *the zeros of* $p_n(x)$ *are distinct and lie in the open interval* (a, b).
2) *if* $x_0 = a$, $x_{n+1} = b$ *and* $x_1, \ldots, x_n$ *are the* n *distinct zeros of* $p_n(x)$*, then each interval* (x_k, x_{k+1}) *contains exactly one zero of* $p_{n+1}(x)$ *for* $k = 0, 1, \ldots, n$.

Proof: We shall only prove (1). The proof of (2) uses a result to be given below (Corollary 9.8) and will not be given (see Exercise 9.7).

If $n \geq 1$, then

$$\int_a^b p_n(x)w(x)\,dx = 0\,,$$

since $p_0(x)$ is a constant. Since $w(x)$ is positive on $[a, b]$ we see that $p_n(x)$ must change sign at least once on the interval $[a, b]$. Let $a_1, \ldots, a_s$ be the points in (a, b) where $p_n(x)$ changes sign. Then the a's are the zeros of $p_n(x)$ of odd multiplicity. Since $p_n(x)$ is a polynomial we must have $s \leq n$. Let

$$q(x) = \prod_{k=1}^{s} (x - a_s) .$$

If $s < n$, then, by Corollary 9.3, we have

$$\int_a^b q(x) p_n(x) w(x) \, dx = 0 .$$

Since the a's are the zeros of odd multiplicity of $p_n(x)$ we must have that $q(x)p_n(x)$ does not change sign on (a, b). Thus we must have $s = n$, that is, $p_n(x)$ has n roots of odd multiplicity in (a, b) and hence, since $p_n(x)$ is a polynomial of degree n and so has exactly n roots, counting multiplicity, all the roots of $p_n(x)$ are distinct and lie in (a, b).

The following theorem gives a recursion relation that allows us to calculate the orthogonal polynomials without having to go through the integral definition (Definition 9.2) providing we have calculated $p_0(x)$ and $p_1(x)$.

Theorem 9.6: *There exist sequences of constants* $\{A_n\}$, $\{B_n\}$ *and* $\{C_n\}$, *with* $A_n > 0$ *and* $C_n < 0$ *for all* n, *such that*

$$p_n(x) = (A_n x + B_n) p_{n-1}(x) + C_n p_{n-2}(x)$$

for $n = 2, 3, \ldots$ *In fact, we have*

$$A_n = \frac{k_n}{k_{n-1}}, \quad B_n = \left(\frac{k'_n}{k_n} - \frac{k'_{n-1}}{k_{n-1}} \right) A_n \quad \text{and}$$

$$C_n = \frac{A_n}{A_{n-1}} = \frac{k_n k_{n-2}}{k_{n-1}^2} \cdot \frac{g_n}{g_{n-1}}$$

where k_m is the leading coefficient of $p_m(x)$ and k'_m is the coefficient of x^{m-1} and g_m is defined by (9.1).

The proof of this result is not very complicated, but is quite computational. Therefore we shall omit the proof, but note that it heavily uses Theorem 9.1 and Corollary 9.3.

As examples of this theorem we give the recursion relation for some of the orthogonal polynomials given below.

1. For the Chebychev polynomials of the first kind $T_n(x)$, we have

$$T_{n+1}(x) = 2xT_n(x) - T_{n-1}(x)\,,$$

with $T_0(x) = 1$ and $T_1(x) = x$.

2. For the Legendre polynomial, $p_n(x)$, we have

$$p_{n+1}(x) = \frac{2n+1}{n+1}p_n(x) - \frac{n}{n+1}p_n(x)\,.$$

3. For the Laguerre polynomials, $L_n^{(\alpha)}(x)$, we have

$$L_{n+1}^{\alpha}(x) = \left(\frac{2n+\alpha+1}{n+1} - \frac{1}{n+1}x\right) L_n^{(\alpha)}(x) - \frac{n+\alpha}{n+1}c_{n+1}^{(\alpha)}(x)\,.$$

4. For the Hermite polynomials, $H_n(x)$, we have

$$H_{n+1}(x) = 2xH_n(x) - 2nH_{n-1}(x)\,.$$

One can use the recursion relation satisfied by the Chebychev polynomials to determine their zeros explicitly. We have

$$T_{n+1}(x) = 2xT_n(x) - T_{n-1}(x)\,,$$

with $T_0(x) = 1$ and $T_1(x) = x$. Suppose we make a change of variables and let $x = \cos\theta$, where $0 \leq \theta \leq \pi$. Then the recursion takes on the form

$$T_{n+1}(\cos\theta) = 2\cos\theta T_n(\cos\theta) - T_{n-1}(\cos\theta)$$

with $T_0(\cos\theta) = 1$ and $T_1(\cos\theta) = \cos\theta$. Thus

$$T_2(\cos\theta) = 2\cos\theta \cdot \cos\theta - 1 = 2\cos^2\theta - 1 = \cos 2\theta$$

and

$$T_3(\cos\theta) = 2\cos\theta \cdot \cos 2\theta - \cos\theta = \cos 3\theta\,.$$

One may then show, by mathematical induction, that

$$T_n(\cos\theta) = \cos(n\theta)$$

or

$$T_n(x) = \cos(n \arccos x)$$

for $1 \leq x \leq 1$. Thus the zeros of $T_n(x)$, all of which lie in the interval $(-1, 1)$ are given by

$$x_{n,m} = \cos\left(\frac{2m-1}{2n}\pi\right),$$

for $m = 1, \ldots, n$ where $x_{n,m}$ is the mth root of $T_n(x)$.

The following theorem and its corollaries are also useful in the theory of Gaussian integration.

Theorem 9.7: *We have*

$$\sum_{j=0}^{n} \frac{p_j(x)p_j(y)}{g_j} = \frac{k_n}{k_{n+1}} \cdot \frac{p_{n+1}(x)p_n(y) - p_n(x)p_{n+1}(x)}{g_n(x-y)}\,.$$

Proof: By Theorem 9.6, we have

$$\begin{aligned}
&p_{n+1}(x)p_n(y) - p_n(x)p_{n+1}(y) \\
&\quad = \left[(A_{n+1}x + B_{n+1})p_n(x) + C_{n+1}p_{n-1}(x)\right]p_n(y) \\
&\qquad - p_n(x)\left[(A_{n+1}y + B_{n+1})p_n(y) + C_{n+1}p_{n-1}(y)\right] \\
&\quad = A_{n+1}(x-y)p_n(x)p_n(y) \\
&\qquad + C_{n+1}\left[p_n(x)p_{n-1}(y) - p_{n-1}(x)p_n(y)\right].
\end{aligned}$$

If we use the values of A_{n+1} and C_{n+1} given by Theorem 9.6, and use mathematical induction the result follows.

If we let $y \to x$ is the result of Theorem 9.7 we obtain the following corollary.

Corollary 9.8: *We have*

$$\sum_{j=0}^{n} \frac{p_j^2(x)}{g_j} = \frac{k_n}{g_n k_{n+1}} \left(p'_{n+1}(x)p_n(x) - p'_n(x)p_{n+1}(x)\right).$$

From this result we immediately obtain the following result.

Corollary 9.9: *We have*

$$p'_{n+1}(x)p_n(x) - p'_n(x)p_{n+1}(x) > 0.$$

From this result we see that if x_k is a zero of $p_n(x)$, then

$$-p_n(x_k)p_{n+1}(x_k) > 0.$$

Since Theorem 9.5(1) states that the zeros of $p_n(x)$ are distinct we see that $p'_n(x_k) \neq 0$. Thus $p_{n+1}(x_k) \neq 0$, and so we see that $p_n(x)$ and $p_{n+1}(x)$ have no overlapping zeros. More than this can be proved and this is the content of Theorem 9.5(2).

We end this section with the following useful theorem, which is a concrete version of Theorem 9.1 for the case of orthogonal polynomials.

Theorem 9.10: *Let $\{p_n(x)\}$ be a set of polynomials orthogonal over $[a, b]$ with respect to the weight function $w(x)$. If $f(x)$ is a polynomial of degree m, then there exist constants $d_0(f), \ldots, d_m(f)$ such that*

$$f(s) = \sum_{k=0}^{m} d_k(f) p_k(x) .$$

Moreover,

$$d_k(f) = \frac{1}{g_k} \int_a^b f(x) p_k(x) w(x) \, dx$$

for $k = 0, 1, \ldots, m$.

Proof: First note that by (9.1),

$$g_k > 0 ,$$

so that each $d_k(f)$ is well-defined.

Since a set of orthogonal polynomials is simple by Definition 9.2, we see, by Theorem 9.1, that there exist constants $d_0(f), \ldots, d_m(f)$ such that

$$f(x) = \sum_{k=0}^{m} d_k(f) p_k(x) .$$

Thus, for $k = 0, 1, \ldots, m$ we have

$$\begin{aligned} \int_a^b f(x) p_k(x) \, dx &= \sum_{j=0}^{m} d_j(f) \int_a^b p_j(x) p_k(x) w(x) \, dx \\ &= d_k(f) \int_a^b p_k^2(x) w(x) \, dx , \end{aligned}$$

by the orthogonality of the polynomials $\{p_n(x)\}$. The result follows from (9.1).

9.2 Gaussian Integration

In this section we shall use some of th results on orthogonal polynomials of the previous section to derive a class of integration formulas that often yield better results than the corresponding Newton-Cote formulas with less computation.

We begin with the key result on this subject.

Theorem 9.11: *Let $\{p_n(x)\}$ be a set of polynomial orthogonal over the interval $[a, b]$ with respect to the positive weight function $w(x)$. Then there exist constants $a_n, \ldots, a_n$ so that*

$$\int_a^b f(x)w(x)\, dx = \sum_{k=1}^{n} a_k f(x_{n,k}) \tag{9.2}$$

is exact for all polynomial of degree less than or equal to $2n - 1$, where the numbers $x_{n,k}$ satisfy

$$x_{n,1} < \cdots < x_{n,n}$$

and are the roots of the nth *degree polynomial $p_n(x)$.*

Proof: Let $f(x)$ be a polynomial of degree at most z_{n-1}. Let $L(x)$ be the Lagrange interpolation polynomial of degree $n-1$ that agrees with $f(x)$ at the points $x_{n,k}$, $k = 1 \ldots, n$. Then

$$L(x) = \sum_{k=1}^{n} \ell_k(x) f(x_{n,k}),$$

where

$$\begin{aligned}\ell_k(x) &= \frac{(x-x_{n,1})\cdots(x-x_{n,k-1})(x-x_{n,k+1})\cdots(x-x_{n,n})}{(x_{n,k}-x_{n,1})\cdots(x_{n,k}-x_{n,k-1})(x_{n,k}-x_{n,k+1})\cdots(x_{n,k}-x_{n,n})} \\ &= \frac{p_n(x)}{p_n'(x_{n,k})(x - x_{n,n})}\end{aligned}$$

for $k = 1, \ldots, n$.

Since $f(x)$ and $L(x)$ are polynomials and agree with each other at the points $x_{n,k}$, $k = 1, \ldots, n$, which are the roots of the polynomial $p_n(x)$, we see that $p_n(x)$ must divide $f(x) - L(x)$. Thus there is a polynomial $r(x)$, of degree at most $2n - 1$, such that

$$f(x) - L(x) = p_n(x)r(x).$$

If we integrate f with respect to $w(x)$ over the interval $[a, b]$, we obtain

$$\begin{aligned}\int_a^b f(x)w(x)\,dx &= \int_a^b L(x)w(x)\,dx + \int_a^b p_n(x)v(x)w(x)\,dx \\ &= \int_a^b L(x)w(x)\,dx \\ &= \sum_{k=1}^{n} f(x_{n,k}) \int_a^b \ell_k(x)w(x)\,dx\end{aligned}$$

since, by Corollary 9.3 in Section 9.1, we have

$$\int_a^b p_n(x)v(x)w(x)\,dx = 0.$$

Thus we may take

$$\begin{aligned}a_k &= \int_a^b \ell_k(x)w(x)\,dx \\ &= \int_a^b \frac{p_n(x)w(x)}{p_n'(x_{n,k})(x - x_{n,k})}\,dx.\end{aligned}$$

Conversely, suppose (9.2) holds for any polynomial f of degree at most $2n - 1$ and some partition points $x_{n,k}$, $k = 1, \ldots, n$. Then it must hold for the polynomial

$$f(x) = \ell(x)v(x),$$

where $\ell(x) = (x - x_{n,1}) \cdots (x - x_{n,n})$ and $v(x)$ in any polynomial of degree at most $n-1$. Then (9.2) implies

$$\begin{aligned}\int_a^b \ell(x)v(x)w(x)\,dx &= \int_a^b f(x)w(x)\,dx \\ &= \sum_{k=1}^{n} a_k f(x_{n,k}) \\ &= 0,\end{aligned}$$

since $\ell(x_n, k) = 0$ for $k = 1, \ldots, n$. Thus, by Theorem 9.2 of Section 9.1, we see that

$$\ell(x) = kp_n(x),$$

for some constant K. Thus, if (9.2) holds for all polynomials of degree at most $2n-1$, then the partition points $x_{n,k}$, $k = 1, \ldots, n$ must be the roots of the nth orthogonal polynomial.

Corollary 9.12: *The weights $\{a_k\}$ given in Theorem 9.11 can also be determined by the following formulas:*

(1)
$$a_k = -\frac{k_{n+1}}{k_n} \cdot \frac{g_n}{p_{n+1}(x_{n,k})p_n'(x_{n,k})},$$

where k_m is the leading coefficient of $p_m(x)$, and

(2)
$$a_k^{-1} = \sum_{j=0}^{n} \frac{p_j^2(x_{n,k}}{g_j}.$$

Proof: To prove (1) let $y = x_{n,k}$ in the formula of Theorem 9.7 of Section 9.1. This yields

$$\sum_{j=0}^{n} \frac{p_j(x)p_j(x_{n,k})}{g_j} = -\frac{k_n}{k_{n+1}} \cdot \frac{p_n(x)p_{n+1}(x_{n,k})}{g_n(x - x_{n,k})}.$$

Thus

$$\begin{aligned}
\sum_{j=0}^{n} \frac{p_j(x_{n,k})}{g_j} \int_a^b p'(x)w(x)\,dx \\
&= -\frac{k_n}{k_{n+1}} \int_a^b \frac{p_n(x)p_{n+1}(x_{n,k})}{g_n(x - x_{n,k})} w(x)\,dx \\
&= -\frac{k_n}{k_{n+1}} \cdot \frac{p_n(x)p_{n+1}(x_{n,k})}{g_n(x - x_{n,k})} \\
&\quad \cdot \int_a^b \frac{p_n(x)}{p_n'(x_{n,k})(x - x_{n,k})} w(x)\,dx \\
&= -\frac{k_n}{k_{n+1}} \cdot \frac{p_{n+1}(x_{n,k})p_n'(x_{n,k})p_k}{g_n},
\end{aligned}$$

by Theorem 9.11.

By the orthogonality of the set of polynomials $\{p_n(x)\}$ we know that if $j \geq 1$

$$\int_a^b p_j(x)w(x)\,dx = 0$$

and if $j = 0$, then $p_0(x)$ is a constant and so we have

$$\sum_{j=0}^{n} \frac{p_j(x_{n,k})}{g_j} \int_a^b p_j(x)w(x)\,dx = \frac{1}{g_0} \int_a^b p_0^2(x)w(x)\,dx = 1.$$

Thus we have

$$1 = -\frac{k_n}{k_{n+1}} \cdot \frac{p_{n+1}(x_{n,k})p_n'(x_{n,k})}{g_n} a_k$$

and the result follows.

To prove (2) let $x = x_{n,k}$ in Corollary 9.8 of Section 9.1. This gives

$$\sum_{j=0}^{k} \frac{p_j^2(x_{n,k})}{g_j} = -\frac{k_n}{k_{n+1}} \cdot \frac{p_{n+1}(x_{n,k})p_n'(x_{n,k})}{g_n} = a_k^{-1},$$

by part (1).

It may be noted that neither of the formulas of Corollary 9.11 seem really suited to calculation of the weights a_k, since the roots of most orthogonal polynomials are not easily calculated numbers in most cases. Fortunately these calculations have been done by others so that there exist tables of $\{a_k\}$ and $\{x_{n,k}\}$ for each commonly used orthogonal polynomial. In the examples that follow we will use the tables given in the *Handbook of Mathematical Functions*, Abramowitz [1]. A more complete compilation can be found in the book by Stroud and Secrest [13].

Before beginning the numerical examples let us calculate the weights for one case, where the calculations are relatively straightforward, namely the Chebychev polynomials of the first kind. Recall that

$$T_n(x) = \cos(n \arccos x)$$

for $n = 0, 1, 2, \ldots$ and that the roots of $T_n(x)$ are

$$x_{n,m} = \cos\left(\frac{(2m-1)\pi}{2n}\right),$$

for $m = 1, \ldots, n$. By (1) of Corollary 9.11 we know that, for $n \geq 1$,

$$a_k = -\frac{k_{n+1}}{k_n} \cdot \frac{g_n}{T_n'(x_{n,k})T_{n+1}(x_{n,k})}.$$

It can be easily shown that

$$g_n = \begin{cases} \pi & \text{if } n = 0 \\ \frac{\pi}{2} & \text{if } n \geq 1 \end{cases} \tag{9.3}$$

(see Exercise 9.8). Also by applying mathematical induction to the recursion formula satisfied by the $T_n(x)$ we see that for $m = 1, 2, \ldots$, $k_m = 2^{m-1}$ since $T_0(x) = 1$ and $T_1(x) = x$ (see Exercise 9.9).

Let $\theta_{n,k} = \frac{2k-1}{2n}\pi$ for $k = 1, \ldots, n$. If we let $\theta = \arccos x$, then $T_n(x) = \cos n\theta$ so that

$$
\begin{aligned}
T_{n+1}(x_{n,k}) &= \cos(n+1)\theta_{n,k} \\
&= \cos\theta_{n,k}\cos n\theta_{n,k} - \sin\theta_{n,k}\sin n \\
&= -\sin\theta_{n,k}\sin n\theta_{n,k}\,.
\end{aligned} \tag{9.4}
$$

It might be noted that

$$
\sin n\theta_{n,k} = \sin\frac{2k-1}{2}\pi = (-1)^k\,. \tag{9.5}
$$

Also

$$
\begin{aligned}
T_n'(x) &= -n\sin n\theta\frac{d\theta}{dx} \\
&= n\sin n\theta\frac{1}{\sqrt{1-x^2}} \\
&= \frac{n\sin n\theta}{\sin\theta}\,.
\end{aligned} \tag{9.6}
$$

Thus, by (9.3), (9.4), (9.5) and (9.6),

$$
\begin{aligned}
a_k &= \frac{-2g_n}{n\dfrac{\sin n\theta_{n,k}}{\sin\theta_{n,k}}\cdot(-\sin n\theta_{n,k}\sin\theta_{n,k})} \\
&= \frac{2g_n}{n} \\
&= \frac{\pi}{n}\,.
\end{aligned} \tag{9.7}
$$

Thus the Gauss-Chebychev quadrature formula is

$$
\int_{-1}^{1} f(x)\frac{dx}{\sqrt{1-x^2}} = \frac{\pi}{n}\sum_{k=1}^{n} f\left(\cos\frac{2k-1}{2n}\pi\right) + E_n\,, \tag{9.8}
$$

where E_n is the error, which will be discussed below.

We give some examples for these integration methods. Except for Example 9.3 on Chebychev integration, where it's not necessary, the weights and partition points for each method are taken from Abramowitz [1].

Example 9.2: Use Gauss-Legendre integration to approximate

$$\int_0^1 \frac{dx}{1+x^2},$$

our standard integral.

Solution. To use Gauss-Legendre integration we must integrate over the interval $[-1,1]$. We could make a change of variable (see below), but it is easier to note that since $\frac{1}{1+x^2}$ is an even function we have

$$\int_0^1 \frac{dx}{1+x^2} = \frac{1}{2}\int_{-1}^1 \frac{dx}{1+x^2}.$$

Thus we take as our function $f(x) = \frac{1}{1+x^2}$. We choose 6 point integration and obtain the value

$$0.785365852\,.$$

Since the exact value is $\arctan 1$, which is 0.785398163, we have an error of 3.231075×10^{-5}. It should be noted that this is better than Romberg integration and uses considerably fewer function evaluations.

Example 9.3: Use Gauss-Chebychev integration to approximate

$$\int_0^1 \frac{\cos x}{\sqrt{1-x^2}}\,dx\,.$$

Solution. As with Gauss-Legendre integration we transform the integral to get the correct integration interval. Since $\frac{\cos x}{\sqrt{1-x^2}}$ is an even function we have

$$\int_0^1 \frac{\cos x}{\sqrt{1-x^2}} = \frac{1}{2}\int_{-1}^1 \frac{\cos x}{\sqrt{1-x^2}}\, dx\,.$$

Thus we take as our function $f(x) = \cos x$. We choose 4 point integration and obtain the value

$$1.201969419\,.$$

Since it is known that the exact value is $\frac{\pi}{2}J_0(1)$, (see Exercise 9.23) where $J_0(x)$ is the Bessel function of order 0, and this is equal to 1.201969714, we see that we have an error of 2.95×10^{-7}.

It should be noted that our original integral had an integrand with a discontinuity at $x = 1$. This is one of the strengths of Gaussian integration, namely that we can sometimes get around the discontinuities of the integrands. In this case the singularities are absorbed into the weight function. If we could have tried a closed Newton-Cotes formula we would not even have been able to start.

Similarly, infinite intervals of integration are difficult to approximate using Newton-Cotes formulas, but using Gaussian integration with Laguerre and Hermite polynomial makes them fairly easy. We shall only discuss the Gauss-Laguerre integration, since Gauss-Hermite integration can be handled in the same way.

Example 9.4: Use Gauss-Laguerre integration to approximate the integral

$$\int_0^{+\infty} \frac{dx}{x^3+1}\,.$$

Solution. To do Gauss-Laguerre integration we need to write our integral in the form

$$\int_0^{+\infty} e^{-x} f(x)\,dx\,,$$

so that for this problem we take

$$f(x) = \frac{e^x}{x^3+1}\,.$$

The extra factor of e^x causes no problem since we will only have a finite sum. Also the tables in Abramowitz [1] give the values of the roots of x_i, the weights w_i and the numbers $e^{x_i} w_i$, so that we will not have to do the extra calculation. We choose 15 point integration and obtain the value

$$1.20898606\,.$$

Since the exact value is $\frac{2\pi}{3\sqrt{3}}$ (see Exercise 9.24) which is 1.209199576, we have an error of -2.1352×10^{-4}.

Another way of handling infinite integrals is to make a change of variable to produce an integral over a finite interval. For example, we write

$$\int_0^{+\infty} \frac{dx}{x^3+1} = \int_0^1 \frac{dx}{x^3+1} + \int_1^{+\infty} \frac{dx}{x^3+1}\,.$$

In the second integral on the right side we make the change of variable $x = \frac{1}{u}$. This gives us the limits 0 and 1 again. (This is why we chose to break the integral at 1: it is its own reciprocal.) Thus, after a little simplification, we have

$$\begin{aligned}
\int_0^{+\infty} \frac{dx}{x^3+1} &= \int_0^1 \frac{dx}{x^3+1} + \int_0^1 \frac{u\,du}{u^3+1} \\
&= \int_0^1 \frac{(1+x)\,dx}{x^3+1} \\
&= \int_0^1 \frac{dx}{1-x+x^2}\,.
\end{aligned}$$

The last integral is easily integrated, in this case, to give the exact value as given above. We could also numerically integrate this using Gauss-Legendre quadrature and in this case we obtain the value 1.20919953, which has an error of -4.6×10^{-8}.

Of course, one might not be so fortunate as to have to do Gauss-Legendre or Gauss-Chebychev over the interval $[0,1]$ or $[-1,1]$. Thus one may have to do a change of variables first. To get from the finite interval $[a,b]$ to the interval $[-1,1]$ we let

$$x = \frac{b-a}{2}t + \frac{b+a}{2}$$

and then we may proceed with the integration. Similar remarks apply to definite intervals of integration. For example, a change of variable shows that

$$\int_{-\infty}^{+\infty} f(x)\,dx = \int_{0}^{+\infty} (f(x) + f(-x))\,dx \tag{9.9}$$

(see Exercise 9.27).

Another remark should be made here as well. Gauss-Chebychev is set up to handle algebraic discontinuities of order $\frac{1}{2}$ at -1 and 1. If we are doing an integral with an algebraic discontinuity of order $\frac{1}{2}$ in the interior of the interval of integration, then we would split the integral up to put the discontinuity at an endpoint. For example, if we wish to approximate the integral

$$\int_0^1 \frac{dx}{\sqrt{x|1-2x|}}$$

we might wish to rewrite it as

$$\begin{aligned}
&\int_0^{\frac{1}{2}} \frac{dx}{\sqrt{x(1-2x)}} + \int_{\frac{1}{2}}^1 \frac{dx}{\sqrt{x(2x-1)}} \\
&\quad = \int_0^{\frac{1}{2}} \frac{dx}{\sqrt{x(1-2x)}} + \int_{\frac{1}{2}}^1 \frac{\sqrt{1-x}}{\sqrt{x}} \cdot \frac{dx}{\sqrt{(1-x)(2x-1)}}.
\end{aligned}$$

The last transformation was made only to keep the problem as one of Gauss-Chebychev integration. By using the other type of Jacobi polynomials, namely the polynomial $p^{(0,\frac{1}{2})}(x)$, we could do the integral

$$\int_{\frac{1}{2}}^{1} \frac{1}{\sqrt{x}} \cdot \frac{dx}{\sqrt{2x-1}}$$

directly, after an appropriate change of variable.

We now discuss error estimates for Gaussian integration. To get the theorem on the error estimate we need two auxilliary results. The first is an extension of Rolle's theorem (Section 2.2), whose proof we leave to the Exercises (see Exercise 9.14). The second result is a corollary of the first and is related to what is known as Hermite interpolation.

Lemma 9.13: *Let $K(x)$ be a function defined on $[a,b]$ that possesses $2n+m-1$ continuous derivatives on $[a,b]$. Suppose $K(x)$ has double zeros on $[a,b]$, at $x_1,\dots,x_n$ and simple zeros on $[a,b]$, at $y_1,\dots,y_m$. Then there exists a ξ in $[a,b]$ such that*

$$K^{2n+m-1}(\xi) = 0\,.$$

Corollary 9.14: *Let $x_1,\dots,n_n$ be distinct values in $[a,b]$ and let $f(x)$ have a derivative of order $2n$ in $[a,b]$. If $H(x)$ is a polynomial of degree $2n-1$ such that*

$$H(x_k) = f(x_k) \quad \textit{and} \quad H'(x_k) = f'(x_k) \quad k = 1,2,\dots,n\,,$$

then there exists a $\xi = \xi(x)$ in $[a,b]$ such that

$$f(x) = H(x) + \frac{f^{(2n)}(\xi)}{(2n)!}(x-x_1)^2\cdots(x-x_n)^2\,.$$

Proof: First of all note that the polynomial H is really unique since it has $2n$ coefficients and there are $2n$ conditions to be satisfied. Once can then show, by modifying the proof of Theorem 5.2 of Section 5.1, that H does exist and is indeed unique.

For $x \neq x_k$ let

$$K(z) = f(z) - H(z) - \frac{f(x) - H(x)}{(x - x_1)^2 \cdots (x - x_n)^2}(z - x_1)^2 \cdots (z - x_n)^2 .$$

Then $x_1, \ldots, x_n$ are double zeros of $K(z)$ and x is a simple zero of $K(z)$. Thus, by the lemma, we have

$$K^{(2n)}(\xi) = 0$$

for some $\xi = \xi(x)$ in $[a, b]$. Since H is a polynomial of degree $2n - 1$ we have $H^{(2n)}(z) = 0$ and it is easily shown that

$$\frac{d^{(2n)}}{dz^{(2n)}}(z - x_1)^2 \cdots (z - x_n)^2 = (2n)! .$$

The result follows.

It should be noted that

$$f^{(2n)}(\xi(x)) = \frac{f(x) - H(x)}{(x - x_1)^2 \cdots (x - x_n)^2}(2n)! ,$$

which is a continuous function of x since f and H at f' and H' agree at the values $x_1, \ldots, x_n$.

We can now state the error formula for Gaussian integration.

Theorem 9.15: *Let $\{p_n(x)\}$ be a set of polynomials orthogonal over the interval $[a, b]$ with respect to the positive weight fuction $w(x)$. Let $x_{n,1}, \ldots, x_{n,n}$ be the roots of $p_n(x)$ on (a, b) and let $a_1, \ldots, a_n$ be the corresponding weights as determined in Theorem 9.11. If f possesses $2n$ continuous derivatives on $[a, b]$, then there exists an η in (a, b) such that*

$$\int_a^b f(x)w(x)\,dx - \sum_{k=1}^{n} a_k f(x_{n,k}) = g_n \frac{f^{(2n)}(\eta)}{(2n)!k_n^2},$$

where

$$g_n = \int_a^b p_n^2(x)w(x)\,dx$$

and k_n is the leading coefficient of $p_n(x)$.

Proof: By the corollary to the lemma above we have

$$w(x)f(x) = w(x)H(x) + \frac{f^{(2n)}(\xi(x))}{(2n)!} w(x) \frac{p_n^2(x)}{k_n^2},$$

where H is the polynomial of degree $2n-1$ determined in the corollary using the points $x_{n,1}, \ldots, x_{n,n}$. Since H is a polynomial of degree $2n-1$ we have, by Theorem 9.11,

$$\begin{aligned}
\int_a^b f(x)w(x)\,dx &= \int_a^b H(x)w(x)\,dx \\
&\qquad + \frac{1}{(2n)!k_n^2} \int_a^b f^{(2n)}(\xi(x))p_n^2(x)w(x)\,dx \\
&= \sum_{k=1}^{n} a_k H(x_{n,k}) + \frac{f^{(2n)}(\eta)}{k_n^2(2n)!} \int_a^b p_n^2(x)w(x)\,dx,
\end{aligned}$$

where we used Theorem 2.8 of Section 2.2 to pull the $f^{(2n)}(\eta)$ out. Since $H(x_{n,k}) = f(x_{n,k})$, by the definition of H, the result follows.

For reference we list the error bounds given by Theorem 9.15 for the Gaussian integration methods we have described. Let

$$E_n(f) = \int_a^b f(x)w(x)\,dx - \sum_{k=1}^{n} a_k f(x_{n,k})\,. \tag{9.10}$$

Then for

1. Legendre polynomials:

$$E_n(f) = \frac{2^{2n+1}(n!)^4}{(2n+1)!((2n)!)^2} f^{(2n)}(\eta)\,, \quad -1 < \eta < 1\,;$$

2. Chebychev polynomials:

$$E_n(f) = \frac{\pi}{(2n)!2^{2n-1}} f^{(2n)}(\eta)\,, \quad -1 < \eta < 1\,;$$

3. Laguerre polynomials:

$$E_n(f) = \frac{(n!)^2}{(2n)!} f^{(2n)}(\eta)\,, \quad 0 < \eta < +\infty\,;$$

and

4. Hermite polynomials:

$$E_n(f) = \frac{n!\sqrt{\pi}}{2^n(2n)!} f^{(2n)}(\eta)\,, \quad -\infty < \eta < +\infty\,.$$

We conclude this section with a result that may seem somewhat remarkable. As the reader will recall from Section 6.4 Newton-Cotes integration does not necessarily produce better results if we increase the number of partition points. Another way of saying this is that there exist functions f, such that the error made between the actual value of the integral of f and its value computed by an n point Newton-Cotes formula does not go to 0 as n goes to infinity. For Gaussian integration this is not the case for finite intervals as is shown in the following theorem due to T.S. Stieltjes, the proof of which uses the Weierstrass Approximation Theorem (Theorem 5.1 of Section 5.1).

Theorem 9.16: *Suppose that f is continuous on the finite interval $[a, b]$. Then*

$$\lim_{n \to +\infty} E_n(f) = 0\,,$$

where $E_n(f)$ is defined in (9.10).

Proof: By (2) of Corollary 9.12 we have

$$a_k > 0 \quad \text{for} \quad k = 1, 2, \ldots, n$$

for any $n \geq 1$. Also if we take $f(x) \equiv 1$, then we have a polynomial of degree 0, and so, by Theorem 9.11, we have, for any $n \geq 1$,

$$\sum_{k=1}^{n} a_k = \int_a^b w(x)\, dx\,,$$

which is independent of n.

Let $\epsilon > 0$. By the Weierstrass Approximation Theorem there is a polynomial P of degree m, say, such that

$$\max_{x \in [a,b]} |f(x) - P(x)| < \frac{\epsilon}{r}\,,$$

where

$$r = 2 \int_a^b w(x)\, dx\,.$$

Suppose $n > m$. Then, by the triangle inequality and Theorem 9.11, we have

$$
\begin{aligned}
|E_n(f)| &= \left| \int_a^b f(x)w(x)\,dx - \int_{k=1}^{n} a_k f(x_{n,k}) \right| \\
&\le \left| \int_a^b f(x)w(x)\,dx - \int_a^b P(x)w(x)\,dx \right| \\
&\qquad + \left| \sum_{k=1}^{n} a_k f(x_{n,k}) - \sum_{k=1}^{n} a_k P(x_{n,k}) \right| \\
&\le \int_a^b |f(x) - P(x)| w(x)\,dx + \sum_{k=1}^{n} a_k |f(x_{n,k}) - P(x_{n,k})| \\
&\le \frac{\epsilon}{r} \left(\int_a^b w(x)\,dx + \sum_{k=1}^{n} a_k \right) \\
&= \epsilon\,.
\end{aligned}
$$

The result follows.

9.3 Least Squares Approximation, II

In this section we shall return to the problem of least squares approximation that we discussed briefly in Section 5.3. There we were concerned with the problem of best fitting a line to a set of data points in the least squares sense. In some of the Exercise Set 5 problems we mentioned how to do a few other types of least squares approximations. In this section we shall discuss the problem of polynomial approximation in two different ways. The first is again to try to fit a polynomial through a set of data points. Of course, we would use the Lagrange or Newton interpolation polynomials to find a polynomial that exactly fits, but if we have lots of points, this is highly impractical. We want to develop a way of fitting a low degree polynomial through many data points. The second problem is to approximate continuous functions over entire intervals in a least squares sense. In both cases it turns out that we again end up using orthogonal polynomials.

Let $\{(x_1, y_1), \ldots, (x_m, y_m)\}$ be a set of data points. We wish to find the polynomial

$$p(x) = a_n x^n + a_{n-1}x^{n-1} + \cdots + a_1 x + a_0$$

that best fits these data points in the least squares sense. That is, we wish to minimize

$$\sum_{k=1}^{m}(y_k - p(x_k))^2 = \sum_{k=1}^{m}\left(y_k - \sum_{j=1}^{n} a_j x_k^j\right)^2 . \tag{9.11}$$

If we proceed as in Section 5.3 (that is, take partial derivatives of (9.11), etc) we find that we must solve the $(n+1) \times (n+1)$ system of equations

$$\sum_{j=0}^{n}\left(\sum_{k=1}^{m} x_k^{j+j}\right) a_j = \sum_{k=1}^{m} y_k x_k^i , \qquad i = 0, 1, \ldots, n .$$

In practice it turns out that this is not the best way to proceed unless n is small, up to about 5, because the system of equations becomes too ill-conditioned for n too large. What one can do is recall that every polynomial can be written as a linear combination of orthogonal polynomials of any one class. For the problem of least squares approximation experience has shown that Chebychev polynomials work best. That is, they give a symmetric matrix of coefficients that is reasonably well-conditioned. Thus, our problem is to now fit constants $a_0, a_1, \ldots, a_n$ so that

$$\sum_{k=1}^{m}\left(y_k - \sum_{j=0}^{n} a_j T_j(x_k)\right)$$

is a minimum.

To use Chebychev polynomials and retain their useful properties we must translate everything to the interval $[-1,1]$. Thus, instead of the data points $\{(x_k, y_k)\}_{k=1}^{m}$ we take the data points $\{(z_k, y_k)\}_{k=1}^{m}$, where

$$z_k = \frac{2x_k - a - b}{b - a},$$

and

$$a = \min_{1\le k\le m} x_k \quad \text{and} \quad b = \max_{1\le k\le n} x_k$$

We then find that the $(n+1)\times(n+1)$ system of equations to solve is

$$\sum_{j=0}^{n}\left(\sum_{k=1}^{m} T_j(z_k)T_j(z_k)\right)a_j = \sum_{k=1}^{m} y_k T_j(z_j), \qquad i = 0,1,\ldots,n.$$

Once we find $a_0, a_1, \ldots, a_n$, by solving this system, we let

$$p(z) = \sum_{j=0}^{n} a_j T_j(z).$$

The polynomial that best fits our original data points in the least squares sense is the polynomial

$$p\left(\frac{2x - a - b}{b - a}\right).$$

We may also note that once we have our best fitting polynomial $p(x)$ if we wish to evaluate it at a particular value of x, there is an easy recurrence scheme to do it. The algorithm is

$$\begin{aligned}
&c_n = a_n, \qquad c_{n-1} = a_{n-1} + 2xc_n, \\
&c_k = a_k + 2xc_{k+1} - c_{k+2}, \qquad k = n-2, n-3, \ldots, 2, 1, \\
&p(x) = a_0 + xc_1 - c_2.
\end{aligned} \tag{9.12}$$

This follows easily from the recurrence relation satisfied by the Chebychev polynomials, namely

$$T_n(x) = 2xT_{n-1}(x) - T_{n-2}(x)\,, \quad n \geq 2\,,$$

$T_0(x) = 1$ and $T_1(x) = x$. We have

$$\begin{aligned}
P(x) &= \sum_{j=0}^{n} a_j T_j(x) \\
&= a_0 T_0(x) + a_n T_n(x) + a_{n-1} T_{n-1}(x) \\
&\quad + \sum_{j=1}^{n-2} (c_j - 2xc_{j+1} + c_{j+2}) T_j(x) \\
&= a_0 T_0(x) + c_n(2xT_{n-1}(x) - T_{n-2}(x)) + a_{n-1} T_{n-1}(x) \\
&\quad + \sum_{j=1}^{n-2} c_j T_j(x) - 2x \sum_{j=2}^{n-1} c_j T_{j-1}(x) + \sum_{j=3}^{n} c_j T_{j-2}(x) \\
&= a_0 T_0(x) + c_{n-1} T_{n-1}(x) - c_n T_{n-2}(x) \\
&\quad + \sum_{j=2}^{n-1} c_j \left[T_j(x) - 2xT_{j-1}(x) + T_{j-2}(x)\right] \\
&\quad - c_{n-1} T_{n-1}(x) - c_2 T_0(x) + c_1 T_1(x) + c_n T_{n-2}(x) \\
&= a_0 T_0(x) - c_2 T_0(x) + c_1 T_1(x) \\
&= a_0 + xc_1 - c_2\,.
\end{aligned}$$

As an example we shall consider a simple case where the data lies in the interval $[-1, 1]$ with both $x = -1$ and $x = 1$ used in the data points. This avoids having to make a change of variables, and so the calculations will not be quite so bad. We take as our data points the following ten points: $(-1, 4.1)$, $(-0.5, 2.184)$, $(0.6, 0.304)$, $(0.3, 0.486)$, $(0.1, 0.81)$, $(0.2, 0.683)$, $(0.35, 0.419)$, $(1, 3.9)$, $(-0.75, 2.848)$, and $(-0.6, 2.436)$.

(These points were obtained by picking the x-values at random in the interval $[-1, 1]$, but including 1 and -1, evaluating the sixth degree polynomial

$$p(x) = 3x^6 + 2x^5 - x^4 - 0.1x^3 + x^2 - 2x + 1$$

and truncating the values obtained to 3 decimal places). The best fitting polynomial of degree 6, in the least squares sense, has the coefficients

$$\begin{aligned} a_0 &= 1.000063 \\ a_1 &= -1.999854 \\ a_2 &= 0.994172 \\ a_3 &= -0.100467 \\ a_4 &= -0.981959 \\ a_5 &= 2.000315 \\ a_6 &= 2.987724\,, \end{aligned}$$

which are not too far off from the correct coefficients considering the truncation used to produce the data points.

We now consider the continuous case. Here we wish to find a polynomial $Q_n(x)$ such that

$$I_n(f) = \int_a^b (f(x) - Q_n(x))^2 w(x)\, dx \tag{9.13}$$

is minimal, where $w(x)$ is a nonnegative, continuous function on the interval $[a, b]$. Let $\{P_n(x)\}$ be the orthogonal polynomials over $[a, b]$ with respect to the weight $w(x)$. Then, by Theorem 9.1 we may write

$$Q_n(x) = \sum_{j=0}^{n} c_j P_j(x)\,. \tag{9.14}$$

Thus, our problem is to find the coefficients c_j, $j = 0, 1, \ldots, n$, as in the discrete case.

If we put (9.14) into (9.13) and expand we get

$$
\begin{aligned}
I_n(f) &= \int_a^b f^2(x)w(x)\,dx \\
&\qquad - 2\sum_{j=0}^{n} c_j \int_a^b f(x)P_j(x)w(x)\,dx \\
&\qquad + \sum_{j=0}^{n}\sum_{\ell=0}^{n} c_j c_\ell \int_a^b P_j(x)P_\ell(x)w(x)\,dx \\
&= \int_a^b f^2(x)w(x)\,dx \\
&\qquad - 2\sum_{j=0}^{n} c_j \int_a^b f(x)P_j(x)w(x)\,dx + \sum_{j=0}^{n} c_j^2 g_j\,,
\end{aligned}
\tag{9.15}
$$

where we have used the orthogonality of the polynomials $\{P_n(x)\}$. A necessary condition for $I_n(f)$ to be minimal is that

$$\frac{\partial I_n(f)}{\partial c_j} = 0\,, \quad j = 0, 1, \ldots, n\,.$$

Since

$$\frac{\partial I_n(f)}{\partial c_j} = -2\int_a^b f(x)P_j(x)w(x)\,dx + 2c_j g_j$$

we see that

$$\frac{\partial I_n(f)}{\partial c_j} = 0$$

if and only if

$$c_j = \frac{1}{g_j}\int_a^b f(x)P_j(x)w(x)\,dx\,. \tag{9.16}$$

To show that (9.16) does indeed yield the minimum for $I_n(f)$ we rewrite (9.14) as

$$
\begin{aligned}
I_n(f) = &\int_a^b f^2(x)w(x)\,dx \\
&-\sum_{j=0}^{n} \frac{1}{g_j}\left(\int_a^b f(x)P_j(x)w(x)\,dx\right)^2 \\
&+\sum_{j=0}^{n}\left(\frac{1}{\sqrt{g_j}}\int_a^b f(x)P_j(x)w(x)\,dx - c_j\sqrt{g_j}\right)^2 .
\end{aligned}
\tag{9.17}
$$

Since $I_n(f) \geq 0$ and the first two terms on the right hand side of (9.17) are independent of the c_j we see that we obtain the unique minimum when the c_j have the values given in (9.16) and with these values we see that the minimum value of $I_n(f)$ is

$$
\int_a^b f^2(x)w(x)\,dx - \sum_{j=0}^{n}\frac{1}{g_j}\left(\int_a^b f(x)P_j(x)w(x)\,dx\right)^2 .
$$

As an example suppose we wish to find the cubic polynomial that gives the least squares approximation to $\cos x$ on the interval $[-1,1]$ with respect to the weight 1. The minimizing polynomial is

$$
\begin{aligned}
Q_n(x) &= a_0 + a_1x + a_2x^2 + a_3x^3 \\
&= c_0P_0(x) + c_1P_1(x) + c_2P_2(x) + c_3P_3(x)\,,
\end{aligned}
$$

where $P_j(x)$, $j = 0,1,2,3$, are the Legendre polynomials and the c_j, $j = 0,1,2,3$ are given by (9.16). It can be shown for the Legendre polynomials

$$
g_n = \frac{2}{2n+1}\,, \qquad n = 0,1,2,\ldots
$$

(see Exercise 9.7). Thus

$$c_0 = \frac{1}{2}\int_{-1}^{1} \cos x \, dx = \sin 1$$

$$c_1 = \frac{3}{2}\int_{-1}^{1} x \cos x \, dx = 0$$

$$c_2 = \frac{5}{2}\int_{-1}^{1} \frac{1}{2}(3x^2 - 1) \cos x \, dx = 15 \cos 1 - 10 \sin 1 \qquad \text{and}$$

$$c_3 = \frac{7}{2}\int_{-1}^{1} \frac{1}{2}(5x^3 - 3x) \cos x \, dx = 0\,.$$

Thus the best cubic is really the quadratic polynomial

$$\begin{aligned} Q_2(x) &= \frac{1}{2}45 \cos(-30 \sin 1)x^2 + 6 \sin 1 - \frac{15}{2} \cos 1 \\ &= -0.46526289x^2 + 0.996458615\,. \end{aligned}$$

By way of comparison the second degree Maclaurin polynomial for $\cos x$ is $p_2(x) = -0.5x^2 + 1$. (Note that this is also the third degree Maclaurin polynomial.) If one were to do further calculations one could see that over the entire interval $[-1, 1]$ $Q_2(x)$ gives a better approximation to $\cos x$ than does the Maclaurin polynomial. Of course, near $x = 0$ the Maclaurin polynomial is excellent, but the farther from the origin we get the better $Q_2(x)$ becomes. For example, at $x = 1$ we have $\cos 1 = 0.540302305$, whereas $Q_2(1) = 0.531295725$, with an error of $9.00658089E - 03$, and $p_2(1) = 0.5$, with an error of 0.040302305.

We mention, in concluding this section, that it is possible to give estimates for the least squares approximants. That is, one can produce results of the form

$$\max_{a \le x \le b} |f(x) - Q_n(x)| \le \epsilon c_n\,,$$

for any $\epsilon > 0$ and $n > n_0(\epsilon)$, where c_n is a constant depending on n. Usually c_n is of the form $n^{-\alpha}$, where α is a positive constant often related to how many derivatives f possesses on the interval $[a, b]$. The proof of such results is beyond the scope of this text as they use further properties of orthogonal polynomials that we have not covered.

9.4 Economization of Power Series

We know that for a sufficiently differentiable function $f(x)$ we can produce polynomial approximations as accurate as we would like from the Taylor series expansion. The trouble with this is that the approximating polynomial we end up with may have a relatively high degree. The aims of economization of power series is to produce an approximating polynomial of smaller degree. This will be achieved through the use of Chebychev polynomials.

Recall from Section 9.1 that the Chebychev polynomials $T_n(x)$, satisfy the following recursion

$$T_{n+1}(x) = 2xT_n(x) - T_{n-1}(x)\,, \tag{9.18}$$

with $T_0(x) = 1$ and $T_1(x) = x$. Thus, for $n \geq 1$ the leading coefficient of $T_n(x)$ is 2^{n-1} (see Exercise 9.9). Also that for $-1 \leq x \leq 1$ we have

$$T_n(x) = \cos(n \arccos x)\,.$$

Thus $|T_n(x)|$, for $-1 \leq x \leq 1$, has a maximum value of 1 which is obtained precisely $n+1$ times at the points $k\frac{\pi}{n}$, $k = 0, 1, \ldots, n$ and that at the successive extreme values $T_n(x)$ alternates in sign. It is this fact and the following theorem which make Chebychev polynomials so useful.

Theorem 9.17: *Among all monic polynomials $P_n(x)$, of degree n, the polynomial $2^{1-n}T_n(x)$ has the smallest upper bound to its magnitude on $[-1, 1]$.*

Proof: By the remarks above we know that for $-1 \le x \le 1$,

$$|2^{1-n}T_n(x)| \le 2^{1-n}.$$

Let $P_n(x)$ be any other monic polynomial of degree n and suppose that

$$\max_{x\in[-1,1]} |P_n(x)| < 2^{1-n}.$$

Since both $P_n(x)$ and $2^{1-n}T_n(x)$ are monic polynomials of degree n we see that $P_{n-1}(x) = P_n(x) - 2^{1-n}T_n(x)$ is a polynomial of degree at most $n-1$.

Now $T_n(x)$ has $n+1$ extreme values on $[-1,1]$ of absolute value 1, so that $2^{1-n}T_n(x)$ has $n+1$ extreme values on $[-1,1]$ of absolute 2^{1-n} and that at these successive extremes $2^{1-n}T_n(x)$ alternate in sign. By hypothesis $|P_n(x)| < 2^{1-n}$ at each of the extreme values. Thus $P_{n-1}(x)$ must change sign at least as many times as there are extreme values of $2^{1-n}T_n(x)$, that is, it must change sign at leat $n+1$ times. Thus $P_{n-1}(x)$, a polynomial of degree at most $n-1$ has at least n roots on $[-1,1]$, which is a contradiction, unless $P_{n-1}(x)$ is identically zero.

Thus we must have

$$\max_{x\in[-1,1]} |P_n(x)| \ge 2^{1-n}.$$

Let $f(x)$ be a given function and suppose we know that on the interval $[-1,1]$ the polynomial $P_n(x)$ approximates $f(x)$ to within an error of E, that is

$$|f(x) - P_n(x)| \le E$$

for all x in $[-1,1]$. Let a_n be the leading coefficient of $P_n(x)$ and let

$$E_{n-1}(x) = P_n(x) - a_n 2^{1-n}T_n(x).$$

Then $E_{n-1}(x)$ is a polynomial of degree at most $n-1$ and we have, for x in $[-1, 1]$,

$$\begin{aligned} |f(x) - E_{n-1}(x)| &\leq |f(x) - P_n(x) + a_n 2^{1-n} T_n(x)| \\ &\leq |f(x) - P_n(x)| + |a_n 2^{1-n} T_n(x)| \qquad (9.19) \\ &\leq E + |a_n| 2^{1-n}\,, \end{aligned}$$

since $|T_n(x)| \leq 1$ on $[-1, 1]$. The polynomial $E_{n-1}(x)$ is called the economized polynomial and its error estimate is given by (9.19).

As an example, consider the Maclaurin expansion for $\sin x$, namely

$$\sin x = \sum_{k=0}^{+\infty} \frac{(-1)^k x^{2k+1}}{(2k+1)!}\,.$$

On the interval $[-1, 1]$ we have

$$\left| \sin x - \sum_{k=0}^{N} \frac{(-1)^k x^{2k+1}}{(2k+1)!} \right| \leq \frac{1}{(2N+3)!}\,,$$

since the series is alternating. Thus the polynomial

$$P_7(x) = x - \frac{x^3}{6} + \frac{x^5}{120} - \frac{x^7}{5040}$$

gives an approximation to within $\frac{1}{9!} < 2.756 \times 10^{-6}$. We wish to compute the economized fifth degree. To do this we will need $T_7(x)$, the seventh Chebychev polynomial, which we can either look up or use the recursion formula (9.18) and the fact that $T_0(x) = 1$ and $T_1(x) = x$. By either method we find that

$$T_7(x) = 64x^7 - 112x^5 + 56x^3 - 7x\,.$$

Thus, the economized polynomial is

$$\begin{aligned} E_5(x) &= P_7(x) - \frac{1}{64 \cdot 5040} T_7(x) \\ &= \frac{46079}{46080} x - \frac{959}{5760} x^3 + \frac{23}{2880} x^5\,. \end{aligned}$$

Thus the total error committed in approximating $\sin x$ on $[-1, 1]$ by $E_5(x)$ is

$$\begin{aligned} |\sin x - E_5(x)| &\leq |\sin x - P_7(x)| + \left| \frac{T_7(x)}{64 \cdot 5040} \right| \\ &\leq \frac{1}{9!} + \frac{1}{64 \cdot 5040} < 5.86E - 06\,, \end{aligned}$$

though as we shall see the error isn't quite this bad since both terms don't reach their maximum at the same place. Finally, for the sake of comparison, we include the fifth degree Taylor polynomial for $\sin x$, namely

$$P_5(x) = x - \frac{x^3}{6} + \frac{x^5}{120}\,.$$

Since $\sin x$, $P_7(x)$, $E_5(x)$ and $P_5(x)$ are all odd we only give their values over the interval $[0, 1]$.

x	$\sin x$	$P_7(x)$	$E_5(x)$	$P_5(x)$
0	0	0	0	0
0.1	0.0998334169	0.099833417	0.0998314166	0.09999987
0.2	0.1986693309	0.198669332	0.1986662708	0.199999556
0.3	0.2955202071	0.295520208	0.2955275833	0.299996628
0.4	0.3894183428	0.389418344	0.3894175416	0.39998578
0.5	0.4794255405	0.479425535	0.4794270833	0.4999566
0.6	0.5646424762	0.56464245	0.5646454792	0.599892006
0.7	0.64421769	0.644217581	0.6442199166	0.699766571
0.8	0.7173560934	0.717355728	0.7173550833	0.799544896
0.9	0.7833269134	0.783325854	0.78332275	0.899179884
1.0	0.8414709871	0.84146826	0.8414713542	0.99861112

Table 9.1

It should be noted that the values of $P_7(x)$ are quite good near $x = 0$. This is because the Maclaurin series is centered around $x = 0$. On the other hand, $P_5(x)$ is a very poor approximation polynomial and gets nowhere near the accuracy obtained by $E_5(x)$. The economization here gives us a polynomial approximant, $E_5(x)$, that gives 5 significant figures and doesn't require as much calculation as $P_7(x)$. One might note that if we economized $E_5(x)$ to a third degree polynomial that the error would be at most $5.05E-04$, which is still better than $P_5(x)$, except very close to $x = 0$.

If we are interested in approximating over an interval $[a, b]$ instead of $[-1, 1]$, then it is a simple matter to make a change of variable from the interval $[a, b]$ to the interval $[-1, 1]$, find the economizing polynomial over $[-1, 1]$ and then make the change of variables back to $[a, b]$.

Exercise Set 9

Mathematical Problems

1. Prove Theorem 9.2 of Section 9.1.

2. Prove the second part of Theorem 9.5 of Section 9.1.

3. Show that if $P_n(x)$ denotes the Legendre polynomial of degree n, then $P_{2k}(x)$ is an even polynomial and $P_{2k+1}(x)$ is an odd polynomial. (HINT: use the recurrence relation.)

4. Show that the Legendre polynomial, $P_n(x)$, satisfies the following Rodriguez formula

$$P_n(x) = \frac{1}{2^n n!} \cdot \frac{d^n}{dx^n}(x^2 - 1)^n .$$

5. Show that if $T_n(x)$ denotes the Chebychev polynomial of the first kind, then $T_{2k}(x)$ is an even polynomial and $T_{2k+1}(x)$ is an odd polynomial.

6. Show that the minimum value of the product

$$\prod_{j=1}^{n}(x - x_j) ,$$

where the x_j are in the interval $[-1, 1]$ for $j = 1, \ldots, n$, is taken on when the x_j are the roots of $T_n(x)$. What is the minimum value?

7. Show that if $P_n(x)$ denotes the Legendre polynomial of degree n, then for $n \geq 0$

$$\int_{-1}^{1} P_n^2(x)\, dx = \frac{2}{2n+1}\,.$$

8. Show that if $T_n(x)$ denotes the Chebychev polynomial of the first kind of degree n, then

$$\int_{-1}^{1} \frac{T_n(x)}{\sqrt{1-x^2}}\, dx = \begin{cases} \pi & \text{if } n = 0 \\ \frac{\pi}{2} & \text{if } n \geq 1 \end{cases}\,.$$

9. Show that the leading coefficients of $T_n(x)$ is 2^{n-1}, $n \geq 1$.

10. Show that $T_n(T_m(x)) = T_{mn}(x)$.

11. If a and b are fixed constants, define the sequence of polynomials $g_k(x)$ by

$$g_k(x) = T_k(ax + b)\,.$$

What recurrence relation do the $g_k(x)$ satisfy?

12. Show that the Laguerre polynomials $L_n^{(\alpha)}(x)$ satisfy the Rodriquez formula

$$L_n^{(\alpha)}(x) = \frac{1}{n!} x^{-\alpha} e^{x} \frac{d^n}{dx^n} x^{n+\alpha} e^{-x}\,.$$

13. If $\{P_n(x)\}$ are orthogonal over $[a,b]$ with respect to the weight $w(x)$, the a_k, $k = 1,2,\ldots,n$, are the corresponding weights used in Gaussian integration and $\{x_{n,k}\}$, $k = 1,2,\ldots,n$, are the distinct roots of $P_n(x)$ on (a,b), then

$$\sum_{k=1}^{n} a_k P_r(x_{n,k}) P_s(x_{n,k}) = 0$$

if $r \neq s$ and $0 \leq r,\ s \leq n-1$.

14. Prove the lemma to Theorem 9.5 of Section 9.2.

15. Prove the following generalization of Rolle's Theorem of which the result of Exercise 14 is a special case.

Theorem: *Let $K(x)$ be a function defined on $[a,b]$. Suppose $x_1,\ldots,x_m$ are the zeros of $K(x)$ on $[a,b]$ and that x_k is a zero of multiplicity n_k. If K possesses derivatives of order up to $n_1+\cdots+n_m$ on (a,b), then there is a ξ in $[a,b]$ such that $\min\{x_1,\ldots,x_m\} < \xi < \max\{x_1,\ldots,x_m\}$ and*

$$K^{((n_1+\cdots+n_m)-1)}(\xi) = 0\,.$$

16. Show that the first degree polynomial one obtains using Chebychev polynomials in a discrete least squares problem is identical with the linear least squares function one obtains by the procedure of Section 5.3.

Computational Problems

17. Compute $P_k(x)$, $T_k(x)$, $L_k(x)$ and $H_k(x)$, for $k = 0,1,2,3,4,$ and 5, using the appropriate recursion formula.

18. Compute $T_k(x)$, for $k = 0,1,2,\ldots,10$ and $x = -1,\ -0.9,\ \ldots,$ 0.9, 1, using the recurrence relation.

For Exercises 19–28 on Gaussian integration refer to the *Handbook of Mathematical Functions* (see Abramowitz [10]) for the appropriate weights and roots. For the exact values of some of these integrals the reader may wish to refer to a table of integrals, for example, that by Gradshtyen and Ryzhik [8].

19. Use 16 point Gauss-Legendre integration to calculate the values of the following integrals.

    a) $\displaystyle\int_0^1 \arctan x \, dx$

    b) $\displaystyle\int_1^2 \cos(\log x) \, dx$

    c) $\displaystyle\int_0^1 \frac{dx}{e^x + e^{-x}}$

    Compare the approximations obtained with the exact values as well as the values obtained for the integrals in the various problems in Chapter 6.

20. Use 16 point Gauss-Legendre integration to compute the integral

$$\int_{1.5}^{3.6} e^x \, dx$$

    and compare this approximation with the exact value.

21. Use Gauss-Legendre integration to calculate the value of the integral

$$\int_0^\pi e^x \cos x \, dx \, .$$

    Use $n = 2, 4, 6, 8, 10, 16$, and 32 points. Compare each of the values obtained with the exact value.

22. Use Gauss-Chebychev integration with 10 points to calculate the value of the integral

$$\int_{-1}^{1} \frac{\ln|x|}{\sqrt{1-x^2}}\, dx\,.$$

Compare the value obtained with the exact value.

23. a) Recall, from Exercise 6.21 of Chapter 6, that

$$J_n(x) = \frac{1}{\pi}\int_0^{\pi} \cos(x\sin\theta - n\theta)\, d\theta\,,$$

where $J_n(x)$ is the $n$the order Bessel function. Show that the following results are valid:

i) $\displaystyle\int_0^1 \frac{\cos(ax)}{\sqrt{1-x^2}}\, dx = \frac{\pi}{2} J_0(a)$

and

ii) $\displaystyle\int_0^1 \frac{\sin(ax)}{\sqrt{1-x^2}}\, dx = \frac{\pi}{2} J_1(a)$.

b) Use 10 point Gauss-Chebychev integration to compute the value of

$$\int_0^1 \frac{x\sin x}{\sqrt{1-x^2}}\, dx\,.$$

Compare the value obtained with the exact value.

24. Show that

$$\int_0^{+\infty} \frac{dx}{1+x^3} = \frac{2\pi}{3\sqrt{3}}\,.$$

25. Use 15 point Gauss-Laguerre integration to compute the values of the integrals:

a) $\displaystyle\int_0^{+\infty} \frac{\sin x}{\sqrt{x}}\, dx$

b) $\displaystyle\int_0^{+\infty} e^{-x}(1-\cos x)\frac{dx}{x}$

c) $\displaystyle\int_0^{+\infty} \frac{\sin x}{x}\, dx$

d) $\displaystyle\int_0^{+\infty} e^{-\pi x^2} \sin^2(2x)\, dx.$

Compare the values obtained with the exact values.

26. Use 15 point Gauss-Laguerre integration to compute the integrals

a) $\displaystyle\int_1^{+\infty} \frac{dx}{x^2}$

b) $\displaystyle\int_1^{+\infty} \frac{\cos x}{\sqrt{x-1}}\, dx$

c) $\displaystyle\int_0^{+\infty} \frac{\ln x}{(1+x^2)^2}\, dx.$

Compare the values obtained with the exact values.

27. Prove equation (9.9) of Section 9.2.

28. Use 15 point Gauss-Laguerre integration to compute the integrals:

    a) $\displaystyle\int_{-\infty}^{+\infty} \frac{dx}{x^2+1}$

    b) $\displaystyle\int_{-\infty}^{+\infty} e^{-x^2} \cos \pi(x+1)\, dx.$

    Compare the values obtained with the exact values.

29. Find a polynomial of degree 3 that best fits, in the least squares sense, the set of data $\{(1.5, \sqrt{1.5}),\ (1,6, \sqrt{1.6}),\ (1.7, \sqrt{1.7}),$ $(1.8, \sqrt{1.8}),\ (1.9, \sqrt{1.9}),\ (2, \sqrt{2})\}$.

30. Find the linear and quadratic polynomials that best fit e^x, in the least squares sense, over the interval $[-1, 1]$ with respect to the weight 1. Compare these polynomials with the Maclaurin polynomials of degrees 1 and 2 for e^x over the interval $[-1, 1]$ at steps of 0.1.

31. Find the linear and quadratic polynomials that best fit e^x, in the least squares sense, over the interval $[-1, 1]$ with respect to the weight $\frac{1}{\sqrt{1-x^2}}$.

32. For $f(x) = \sin \frac{\pi x}{2}$, $-1 \le x \le 1$, find both the Legendre (weight $w(x) = 1$) and Chebychev (weight $w(x) = \frac{1}{\sqrt{1-x^2}}$) least squares polynomial approximation of degree 3 over $[-1, 1]$. Compare the values of these polynomials over $[-1, 1]$ at steps of 0.1.

33. Find the Maclaurin polynomial that approximates e^x to within 10^{-5} on $[-1, 1]$. Find the economized polynomials of degrees one less and two less. Compare the values over $[-1, 1]$, at steps of 0.1, of e^x and the three polynomials.

34. Recall from Exercise 2.7 of Chapter 2 that

$$J_0(x) = \sum_{k=0}^{+\infty} \frac{(-1)^k (x/2)^{2k}}{(k!)^2}.$$

Find the value of n so that the partial sum from $k = 0$ to n approximates $J_0(x)$ with an error of less than 10^{-5} over the interval $[-1, 1]$. Find the economized polynomials of degree one less. Compare the values over $[-1, 1]$, in steps of 0.1, of $J_0(x)$ and the two polynomials. You may need to consult tables to find the values of $J_0(x)$.

References

1. Abramowitz, M., and Stegan, I., eds., *Handbook of Mathematical Functions with Formulas, Graphs and Mathematical Tables*, Dover, 1960.
2. Atkinson, K.E., *Introduction to Numerical Analysis*, Wiley, 1978.
3. Bartle, R.G., *Elements of Real Analysis*, Wiley, 1976.
4. Berg, P.W., and McGregor, J.L., *Elementary Partial Differential Equations*, Holden–Day, 1966.
5. Boyce, W.E., and Prima, R.C.D., *Elementary Differential Equations*, Wiley, 1977.
6. Burden, R.L., Faires, J.D., and Reynolds, A.C., *Numerical Analysis*, PWS, 1981.
7. Davis, P.J., and Rebinowitz, P., *Methods of Numerical Integration*, Academic Press, 1984.
8. Gradshteyn, I.S., and Ryzhik, I.M., *Tables of Integrals, Series and Products*, Academic Press, 1980.
9. Michens, R.E., *Difference Equations*, Van Nostrand–Reinhold, 1987.
10. Rainville, E.D., *Special Functions*, Chelsea, 1960.

11. Rivlin, T.J., *Chebyshev Polynomials*, Wiley, 1990.
12. Strang, G., *Linear Algebra and its Applications*, Academic Press, 1976.
13. Stroud, A.H., and Secrest, D.H., *Gaussian Quadrature Formulas*, Prentice–Hall, 1966.
14. Szegö, G., *Orthogonal Polynomials*, AMS, 1975.

Index